T0219339

HARVARD EAST ASIAN MONOGRAPHS

97

China's Silk Trade:
Traditional Industry in the Modern World
1842–1937

China's Silk Trade:
Traditional Industry in the Modern World
1842-1937

LILLIAN M. LI

PUBLISHED BY
COUNCIL ON EAST ASIAN STUDIES
HARVARD UNIVERSITY

Distributed by
Harvard University Press
Cambridge, Massachusetts and London, England
1981

The Council on East Asian Studies at Harvard University
publishes a monograph series and, through the Fairbank
Center for East Asian Research, administers research
projects designed to further scholarly understanding of
China, Japan, Korea, Vietnam, Inner Asia, and adjacent areas.

Library of Congress
Cataloging in Publication Data

Li, Lillian M., 1943–
China's silk trade.

(Harvard East Asian monographs ; 97)
Bibliography: p.
Includes index.
1. Silk manufacture and trade—China—History.
I. Title. II. Series.
HD9926.C62L4   382′.4567739′0951   81–2409
ISBN 0–674–11962–2   AACR2

To My Parents

Contents

Tables

Figures

Maps

Acknowledgments

Teachers, colleagues, friends, and family have given me abundant advice and encouragement during the years when this work has evolved from a would-be dissertation into an actual book. Like so many others, my greatest intellectual debt is to John King Fairbank, who has set the highest academic and personal standards for his students, and has created an academic environment in which both scholars and scholarship can flourish. I am also greatly indebted to Dwight H. Perkins, an economist whose broad tolerance and support of historians is well known and who provided important guidance.

I own special thanks also to Ramon H. Myers and Thomas G. Rawski, two economists who offered incisive criticism in a witty and generous manner. Among others who read all or parts of this work in its various stages and gave suggestions, I would like particularly to thank Francesca Bray, Sherman Cochran,

Albert Feuerwerker, Robert Gardella, Winston Hsieh, Rhoads Murphey, Nathan Sivin, Paul Smith, and E-tu Zen Sun.

The unsung heroes and heroines of all works of Chinese scholarship created at Harvard University are the staff of the Harvard-Yenching Library, whose erudition, helpfulness, and good nature have immensely facilitated the work of this and other scholars. Swarthmore College and the Fairbank Center for East Asian Research at Harvard University have also given me crucial institutional support.

My greatest personal debt is to my parents, who have consistently supported me in my academic pursuits, even when they were justifiably skeptical about their value and outcome. On occasion my father has given me valuable scholarly assistance, and his calligraphy graces the cover of this book.

Cambridge, Massachusetts
June 1980

Note on Transliteration and Names

This book employs the Wade-Giles system of romanization of Chinese names and terms. Most well-known geographic terms, however, follow the conventional pre-1949 postal spellings, for example: Kiangsu instead of Chiang-su, or Tientsin instead of T'ien-chin. For personal names, normal Chinese or Japanese word order is followed, with the surname first, with the exception of those Chinese or Japanese authors who have written for English-language publications, for example Liang-lin Hsiao, instead of Hsiao Liang-lin.

Weights and Measures

Weight
   1 picul (*tan*) = 100 catties = approx. 133 lbs. = approx. 60 kg.
   1 catty (*chin*)[a] = 16 Ch. ounces (*liang*)

Length
   1 *li* = .40057 miles = .6464 km.
   1 Ch. foot (*ch'ih*) = 14.1 inches = 35.81 cm.

Area
   1 mou[a] = approx. 1/6 acre

Money
   1 Haikwan (HK) tael (*liang*) = approx. 1.50 dollar (yuan)[b]
   1 cash (*ch'ien*) = .001 tael

Notes: [a]For the relationship between *chin* and mou, and the *shih-chin* and *shih-mou* imposed after 1928, see Table 1, note a.

[b]For the exchange rate between Haikwan tael and other currencies, see Tuan-lin Yang and Hou-pei Hou, *Statistics of China's Foreign Trade during the Last Sixty-five Years, 1864–1928* (Nanking, 1931), p. 151.

xv

Introduction

Of all the products associated with the material wealth and cultural splendor of traditional Chinese civilization, none was so quintessentially Chinese as silk. From the most ancient times silk played a role in Chinese history, both as a symbol of imperial tradition and as a mainstay of the peasant economy. Archaeological evidence shows that sericulture was practiced in prehistoric times.[1] Mythology associates the origin of sericulture with the legendary emperors Fu Hsi and Shen Nung. In historic times, just as it was the duty of the emperor to set an example for the people by ritually plowing the fields, so too the empress was to lead the nation in sericulture. If agriculture was the root of Chinese civilization, sericulture was one of the branches.

Sericulture served the ultimate purpose of providing the court with gorgeous brocades and satins necessary to maintain imperial splendor, but silk was more than a luxury to be enjoyed and flaunted. It was a form of wealth and became an important part of the fiscal system of the Chinese state. For example, under the equal-field system of land tenure, which was practiced in some parts of China from the sixth to the eighth centuries A.D., mulberry fields as well as rice fields were allocated by the government to peasant households, because both sericulture and agriculture were considered to be essential to their livelihood and to the fulfillment of their tax obligations to the state. After the demise of the equal-field system, tax obligations continued to be calculated in grain, silk cloth, and labor services, although often these were commuted into cash payments. During the Ming dynasty (1368–1644), some "farmland silk" was theoretically required from each hsien (county), but in reality the bulk of the government silk came from the main silk region in the Kiangnan area—the area south of the lower Yangtze River comprising the southern prefectures of present-day Kiangsu province and the northern prefectures of Chekiang province.[2] After the "Single-Whip" tax reforms of the late Ming, silk continued to play an indirect role in the tax system. Since taxes were then calculated in silver, peasants in some areas turned to silk handicraft production in order to earn extra income with which to pay their taxes.[3]

While fulfilling the needs of the state, sericulture and silk manufacture also played an important role in the peasant economy of several regions of China. Sericulture originated in north China and was particularly well developed in Shantung during the Ch'in and Han (221 B.C.–220 A.D.).[4] It was not introduced into the south until the period of the Northern and Southern dynasties (420–589). During the Northern Sung (960–1126), it was practiced in all but two of the twenty-three provinces of China, but was increasingly concentrated in three major regions: the middle and lower Yellow River valley, Szechwan, and the middle and lower Yangtze valley. Although the Yangtze valley contributed an increasingly large proportion of the silk paid to the central government as taxes and tribute, the quality of silks manufactured in north China was considered superior.[5] Not until the movement of the Sung capital to south China after 1127 did the lower Yangtze valley begin to establish its reputation as the most important silk region in China. Sericulture in the localities surrounding the capital of Hangchow flourished quickly. Although most of the silk of this area served the needs of the state, either directly or indirectly, the silk trade nevertheless became increasingly commercialized, complex, and specialized.[6]

During the Ming period, Kiangnan began to eclipse its major competitors, the Shantung-Hopei region and Szechwan province, as the leading silk center of China. In part this reflected an increasing regional specialization. When cotton became popular in China during the Yuan dynasty (1260–1368), it displaced silk and hemp in some of their uses. No longer was it the case that households all over China needed to raise silkworms for their own use.[7] Since the Kiangnan region was naturally suited to sericulture, the Imperial Silk Weaving Factories became centered there. By the early Ch'ing dynasty (1644–1911), only the factories at Hangchow, Soochow, and Nanking were still of importance for the court, while those of outlying districts and provinces had been largely abandoned. Sericulture and silk manufacture were an important aspect of the increasing agricultural productivity and the growing commercialization of the Kiangnan region in the late Ming and early Ch'ing period. The contribution of sericulture to the wealth of these localities was legendary. The journalist Wang T'ao observed in his travels in this area in the 1850s, "From Chia-hsing to P'ing-hu there are mulberry trees planted all along the river. If you raise silkworms and produce silk, your income will increase a hundred times. This is really a source of wealth for this area."[8]

Silk was as important to China's foreign relations as it was to its domestic economy. From the earliest contacts between China and the West, silk played an important role both as a commodity of trade and a medium of exchange. During the former Han period (206 B.C.–8 A.D.), there is archaeological evidence that silk was used as currency when Han envoys and merchants traveled in Central Asia.[9] In fact, in the ancient Roman Empire, silk was identified by the Greek word for the Chinese, "Seres." Until the modern era, the desire to acquire Chinese silk was one of the major forces drawing Western merchants to China. The Chinese silk used in the Roman Empire was imported by Parthian middlemen who traversed the Silk Road from Ch'ang-an through Central Asia. According to legend, during the sixth century A.D., two monks smuggled silkworms from China to Europe inside their bamboo canes and thus introduced sericulture to the Western world.[10] Promoted by the Byzantine emperor, Justinian I, sericulture spread throughout Syria and Greece. Later the Arabs spread it to Sicily and Spain, whence it was introduced by the Normans to Italy. Despite the European adoption of sericulture, it is believed that Chinese silk continued to be exported to Europe during the Middle Ages because of its higher quality. During the time of the Mongol Empire, when trade between East and West was greatly facilitated, silk was the principal commodity traded.[11]

Silk was also the key commodity in China's trade and tributary relations with Central Asian neighbors. When the embassies of foreign countries traveled to the Chinese capital to pay homage to the Chinese emperor, they received in return for their costly tribute offerings, valuable Chinese gifts to bear back to their own monarchs. Silk was probably the most treasured of these commodities. From the Hsiung-nu in the Han period to the Mongols in the Ming period, all the "barbarian" states received gifts of silk, sometimes in exchange for much needed commodities such as horses. Thousands of bolts of satins and brocades thus formed the medium of traditional Chinese diplomacy.[12]

Silk was also a major commodity in China's overseas trade prior to the Treaty of Nanking in 1842. During the late Ming and the early Ch'ing, substantial quantities of Chinese silk were exported to Southeast Asia, Japan, and Spanish America. The Portuguese and the Spanish, having established themselves at

Macao and Manila respectively in the sixteenth century, became the middlemen in the lucrative Japanese and Spanish American trades. Silk was also the major commodity shipped from China by the British East India Company until the mid-eighteenth century, when tea became almost the raison d'être of British trade in Asia.

It was not until the opening of the treaty ports, however, that silk exports to the West reached a large scale. Chinese silk fulfilled a need created when silkworm disease ruined European sericulture in the mid-nineteenth century. Exports more than doubled between 1868 and 1900. By the turn of the century, silk had surpassed tea as China's leading export and, even with the diversification of China's overseas trade, silk continued to be its major export until the 1930s. Despite this steady growth, China's silk export industry could not successfully compete with the Japanese for a larger share of the rapidly expanding world market in this period. From a much smaller base than China's, Japan had developed during the Meiji period (1868–1911) a flourishing silk industry, which was more scientifically organized and responsive to foreign tastes than its Chinese counterpart. Although the best grades of Chinese raw silk were still preferred by French buyers, most foreign customers preferred Japanese silk because of its standard quality and reliability. American silk manufacturers, who particularly required standardization for their power looms, were the major consumers of raw silk and became the most important customers of the Japanese.

Not until the 1920s did Chinese merchants and the Chinese government, with international encouragement and cooperation, begin to take serious steps toward the promotion and regulation of the silk industry. But it was too late. The decline of the industry was irreversible. The introduction of synthetic fibers after World War I began to provide a cheap and reliable substitute for silk. The rural areas in China that produced silk never recovered from the effects of the Great World Depression. And the occupation of the silk-producing regions of east China by Japanese troops after 1937 dealt the final blow to a once-venerable industry.

Japan's silk export trade was critically important in its economic modernization. In concrete terms, development of a major export commodity enabled Japan to earn the foreign exchange to purchase machinery and raw materials necessary for its industrialization. W. W. Lockwood has estimated that "probably the raw silk trade financed no less than 40 percent of Japan's entire imports of foreign machinery and raw materials used domestically" between 1870 and 1930.[13] In a less concrete but equally important sense, the development of the raw silk industry for export provided Japanese entrepreneurs with an opportunity for profitable investment and an incentive toward standardization and, later, technological modernization. It was, as one economist has said, "a training school for Japanese industrialization."[14]

Although differences in scale would have dictated that foreign trade could never have played as significant a role in the Chinese economy as it did in the Japanese economy, nevertheless, China's failure to take better advantage of the international market for silk does suggest that an opportunity to enrich the country through increased exports was lost. That failure also suggests certain historical paradoxes. The silk industry was one of traditional China's most important industries, both in terms of the state's interests and the peasant's livelihood. Moreover, it was a distinctly "advanced" sector of the economy: its technology was particularly well developed, and its form of production was in many ways more complex than that of other handicrafts. In addition, the importance of sericulture and silk manufacture was universally recognized within China. Yet, even before the modern era, technology seems to have stagnated, albeit at a rather high level. And later, the stimulus of Western trade did not result in quick or widespread technological response. In absolute terms, trade did expand but, when viewed internationally, the Chinese silk industry compared unfavorably with its Japanese rival.

These paradoxes are well known to those familiar with the larger issues of Chinese history. Traditional Chinese institutions and technology looked fine in their own context, but suffered when placed in the global context of the nineteenth and twentieth centuries. The silk industry provides an excellent microcosm of these paradoxes. It permits us to examine in some detail the nature of a traditional enterprise, spanning the continuum from agriculture, through several stages of processing, to its final marketing. Moreover it enables us to see the functioning of an advanced sector of the traditional economy—one in which some Marxists have claimed there were the "sprouts of capitalism." How were production and commerce organized? What was the role of the state in this sector of the economy? Why did technological stagnation follow a period of technological innovation? In short, the silk trade permits us to reexamine some fundamental questions about "science" and "capitalism" in the traditional Chinese economy.

It also permits us to examine how a traditional industry reacted to the opportunities for Western trade after 1842. In discussions of the impact of foreign trade and investment on China, most attention has been focused on the damaging effects of imports on traditional Chinese handicrafts. The dominant Chinese view has been that foreign imports, benefiting from the unequal treaties, quickly captured the domestic market for cotton goods and ruthlessly destroyed the "sprouts of capitalism" in the cotton industry. If it had not been for Western imperialism, these sprouts would have developed into fully mature plants. But recent Western scholarship has tended to argue that the economic impact of imperialism was largely restricted to the treaty ports and did not penetrate inland.[15] Moreover, in the case of cotton, some have argued that the impact of foreign trade and investment affected different sectors of the handicraft industry in different ways, not all of them harmful.[16]

The silk industry permits us to examine the impact of foreign trade from the export side. Rather than suffering under a competitive disadvantage, the silk industry benefited from considerable advantages—not only its previous experience and technological skill but also an expanding world market. But in order for the Chinese silk industry to maintain its advantage in the relatively free world market, it had to "modernize." Although "modernization" can be defined in narrow technological terms, this case study shows that the failure to modernize rapidly lay, not in any deficiency of technological potential, but in the commercial practices and institutional arrangements of the silk business, and in an economic environment that provided insufficient incentives. This is not to say that the silk industry failed to "respond" to the world market. Indeed the industry was greatly stimulated by Western demand, and the expansion of exports was impressive. Simply, the Chinese response did not take the form of rapid modernization, as it did with the Japanese. In short, the silk industry permits us to see, with a fair degree of accuracy, those aspects of traditional Chinese economic organization that were amenable to externally stimulated change, and those that were not. Moreover it permits us to separate, for analytical purposes, the question of "modernization" from the question of "imperialism."

During the late nineteenth and early twentieth centuries, the opportunities for exporting silk stimulated the development of sericulture in several regions of China, most notably in Kiangnan, Kwangtung, and Shantung but, since it is the relationship between the traditional silk industry and the new export market which forms the central focus of this study, the Kiangnan region receives the most detailed analysis. In any case, Kiangnan is of the greatest importance for a study of the modern export trade because its white raw silk, exported through Shanghai, was China's premium quality silk, the type most desired in France and the United States. Kiangnan was not China, however, and its experience is contrasted with that of other areas where relevant. At all times this study is concerned with the significance of the silk export trade for China's economy as a whole and its economic institutions in general.

ONE

The Technology of Silk

Chinese have long prized silk fabric for its softness, luster, and affinity for brilliant dyes. Although silk garments were worn mostly by the wealthy, they were not beyond the reach of ordinary people for special attire on ceremonial occasions such as weddings. The manufacture of silk is the result of a perfect union of natural miracle and human intelligence. In feeding on mulberry leaves, passing through four molting periods, and finally spinning a cocoon of silk fiber, the silkworm enacts a complete biological process. But without human attention and skill during this process, the silkworm would produce only wild silk, which is unsuitable for weaving into fine fabrics. Silk manufacture can be divided into four basic stages: the cultivation of mulberry trees, the raising of silkworms (sericulture), the reeling of silk fiber from cocoons, and the weaving of silk fabrics. In each of these stages, Chinese technology had in antiquity reached a level of skill unsurpassed in the world and, by the seventeenth century, the particular techniques of the Kiangnan region had reached a level unsurpassed in China. The purpose of this chapter is to examine this traditional technology, with particular reference to Kiangnan, in order to identify in what sense it was advanced, and in which ways it was, or could have been, improved by modern innovations.

MULBERRY TREES

Although silkworms will eat the leaves of various types of trees, only if they are fed the leaves of the mulberry tree will they produce silk of high quality. In China sometimes leaves from the *che* tree were fed to silkworms when mulberry leaves were scarce, but the cocoons produced were small and yielded much less silk.[1] In Shantung and Manchuria there were wild silkworms which fed on oak leaves, but their silk was quite coarse. Thus, the manufacture of fine silk could flourish only where mulberry cultivation was well developed. There are many varieties of the mulberry, which belongs to the Moraceae family, but in the Kiangnan area only the white mulberry, *morus alba,* was grown. Traditionally, it was raised to produce two crops of leaves annually, the main one in the spring and a smaller one in the summer. By contrast, the *morus latifolia,* grown in south China, could produce enough leaves for seven crops of silkworms a year.[2]

Mulberry trees are hardy and can be grown in almost any type of soil and climate.[3] In China it was widely held that mulberry could be grown anywhere except in low flooded fields.[4] The sericultural manuals said that there was no land suitable for grain which was not also suitable for mulberry. However, in order for the trees to be highly productive, the ground had to have an adequate system of drainage and be well fertilized.[5]

The loamy quality of soil in northern Chekiang was particularly suitable for mulberry, but it was not a necessary condition for its cultivation.[6]

Since mulberry trees were not soil-specific, they could be planted on slopes or gravelly areas not suitable for annual crops, but ideal for trees because of their good drainage.[7] In Kiangnan, mulberry trees were planted on hills, along the banks of canals, between rice paddies, along walls, wherever there was space.[8] One late Ch'ing manual said, "In every household, in every village, there is not one place without some crevice in the wall or land. Aside from walking paths, there is no place where you cannot grow mulberry trees."[9] Traveling near Hangchow in the 1850s, the English botanist Robert Fortune observed, "The appearance of the flat country here was rich and beautiful. Still the mulberry was seen extensively cultivated on all the higher patches of ground, and rice occupied the low, wet land."[10] Fields devoted exclusively to mulberry cultivation were extremely rare, limited to a few highly specialized localities in Hu-chou or Chia-hsing prefectures.[11] In other parts of Kiangnan, various vegetables, and even wheat, were planted in the same field; only bean sprouts, mustard plants, and barley were not supposed to be planted with mulberry.[12] In some areas, even cotton was planted with mulberry saplings.[13] In most provinces, marginal planting and mixed planting were the rule.[14]

In Kiangnan, there were basically two kinds of mulberry trees, the *Lu-sang* and the *Ching-sang*. When a branch of the *Lu-sang* was grafted onto the trunk of the *Ching-sang*, the large leaves of the former were combined with the strong trunk and roots of the latter.[15] Since the *Lu-sang* was almost always grafted, it was known colloquially as *chia-sang,* or domestic mulberry, while the *Ching-sang* was known as *yeh-sang,* or wild mulberry. Because the *Lu-sang* from the Hu-chou area of northern Chekiang was the most widely used in Kiangnan, it was known as the *Hu-sang.* In the nineteenth and twentieth centuries, the two terms were used interchangeably.[16] The *Hu-sang* was generally started from seed, grafted, and then transplanted in the third year as a sapling to the mulberry field of the buyer. After another two or three years, its leaves could be picked for feeding to silkworms. Although the propagation of mulberry trees through layering—pressing a branch of a tree into the ground and permitting it to root—had been more common in earlier times, by the Ch'ing period the sapling method was widespread because it saved the peasant two or three years of time, despite the risks involved in the transplanting process.[17]

Once transplanted, the most critical factors in the productivity of the tree were the application of fertilizer, cultivation of the soil, and pruning. Various types of fertilizers could be used—night soil, animal manure, river mud, beancakes, and so on. One of the reasons mulberry grew so well in Kiangnan was that the dense network of waterways provided a plentiful supply of river mud, which was among the best fertilizers. The seventeenth-century agricultural manual *Pu Nung-shu* said, "If the family does not thrive, it is for the lack of attention; if the cultivation of mulberry does not thrive, it is for the lack of river mud."[18] Nineteenth and twentieth century sources also emphasized the benefits of river mud.[19] Regardless of the type of fertilizer, it was widely recognized that there was a direct correlation between the amount used and the yield of the trees. In his manual *Shen-shih Nung-shu,* on which *Pu Nung-shu* was based, the author attributed his success in raising mulberry trees to the generous use of fertilizer. He cited an old expression, "In raising mulberry trees, one cash more fertilizer means two cash more leaves."[20]

Cultivation of the soil was also essential. Robert Fortune observed that "the Chinese seem very particular in stirring up the earth amongst the roots of the bushes immediately after the young branches have been taken off, and the plantations appear to have great attention paid to them."[21] The mulberry trees also had to be pruned regularly to produce more leaves. Although trees were allowed to grow taller elsewhere (see Figure 1), in Chekiang, the trees were usually pruned to a round shape and were not allowed to grow more than four to six feet high so

FIGURE 1 Picking Mulberry Leaves

Source for Figures 1–4: Yü-chih keng-chih t'u, comp. Chiao Ping-chen, 1696 ed.

that the leaves could be picked without using a ladder.[22] North of the T'ai-hu, in Kiangsu, the trees were only about two or three feet tall.[23] Finally, the proper care of trees also required constant vigilance against insects and worms. Mr. Shen recommended scraping the pests off the trees at least three times a year.[24]

Although the authors of Ch'ing agricultural manuals and twentieth-century observers alike emphasized the extent of local variations in such techniques,[25] what seems clear with the benefit of hindsight is the essential consistency of practices from at least the seventeenth century to the twentieth. The only discernible difference in mulberry cultivation between the early Ch'ing and the twentieth century seems to lie not in the technology employed, but in the results achieved. The various figures on yields culled from these sources, as summarized in Table 1, suggest that mulberry yields probably declined from the early Ch'ing to the twentieth century. The *Pu Nung-shu* states that the yield of mulberry trees could vary from 800 to 2,000 catties (*chin*) per mou. The yield from trees planted in scattered areas could be as high as 4,000–6,000 catties per mou. Figures given by later Ch'ing sources tend to give a narrower and more conservative range of figures, from 1,000–1,500 catties per mou. Data from the 1920s reveal a decline, at least in the upper range of yields. Several sources agree that 1,200 catties was the absolute maximum yield for Chekiang, and 1,000 catties was the maximum for Kiangsu.

These data must be treated with a fair amount of skepticism. The yield of mulberry trees was highly variable, depending on a wide variety of factors such as the age and height of the trees, the density of planting, and the number of crops picked per year, and consequently difficult to measure with any accuracy. The fact that most mulberry planting was done in marginal or mixed areas would have made yields even more difficult to calculate. Nevertheless, it does not seem surprising that, as sericulture spread beyond the northern Chekiang area, the intensive, garden-like methods employed there in the early

TABLE 1 Mulberry Yields in Kiangnan

Date	Place	*Annual Yield per Mou*[a]	
		chin/mou	*shih-chin/ shih-mou*
17th century	Chia-hsing	800–2,000[b] (enclosed)	800–2,800
16th–19th centuries	Chia-hsing	4,000–6,000[c] (scattered)	4,998–7,498
16th–19th centuries	Chia-hsing	800–1,800[d]	999–2,249
16th–19th centuries	Hu-chou	1,600[e]	1,999
18th–20th centuries	Hu-chou	1,300–1,400[f]	1,624–1,749
19th century	Tan-t'u, Chinkiang	1,200[g]	1,499
1910s	Hu-chou	1,000[h]	1,249
1910s	Hangchow	500–1,300[i]	622–1,624
1920s	Hangchow, Chia-hsing	1,000–1,200[j]	1,249–1,499
1920s	Soochow, Wusih	800–1,000[j]	1,000–1,249
1930s	Wu-hsing	–	1,000[k]
1930s	Chekiang & Kiangsu	–	700–1,500[l]
1930s	Wusih	663[m]	789
1956	Wu-hsing, T'ung-hsiang	–	178–815[n]

Notes: [a]Calculations from old *chin* (catties) per old mou to *shih-chin* per *shih-mou* were made according to the formula (*chin* x 1.19) 1.05 = *shih-chin/shih-mou*. This formula is based on Ch'en Heng-li's calculation that one old mou = 0.95 *shih-mou*, and one old *chin* = 1.19 *shih-chin*. See Ch'en Heng-li, *Pu Nung-shu yen-chiu* (Peking, 1958), pp. 36–37, where this formula is used, and pp. 293–300, where these equivalents are developed. Wu Ch'eng-lo, *Chung-kuo tu-liang heng shih* (Shanghai, 1937), pp. 60–61, confirms the latter figure, but declines to make a definite statement on the mou. The conversion from old *chin* to *shih-chin* was part of the Nationalist Government's move to standardize weights and measures after 1928. Ibid., p. 233.

[b]Ch'en Hengli, pp. 36–37. The *Pu Nung-shu* said that land of ordinary variety would produce 40–50 *ko* per mou, while the best grades of land could produce

80-90 *ko* per mou. Since 1 *ko* = 20 *chin*, it follows that yields could range from 800 to 1,800 *chin* per mou, and even to 2,000 *chin* per mou, according to Ch'en.
cCh'en Heng-li, pp. 38-39. *Hu ts'an-shu,* Wang Yueh-chen (1874; Peking reprint, 1956.), p. 20, quotes the *Kuang ts'an-sang shuo* as saying that 100 mulberry trees would yield 20-30 piculs of leaves. (1 picul = 100 catties). This would yield 4,000-6,000 catties, assuming there were 200 trees per mou. This of course was a much higher yield than that cited even in the *Pu Nung-shu.* Ch'en Heng-li remarked that the yield for trees planted in a scattered fashion was much higher than for those planted in an enclosed field.
dChia-hsing FC (1879), 32:24. This passage says, "One mou of land will yield 80-90 *ko*, 20 *chin* being one *ko.* This is double the yield of 40-50 *ko* for medium-grade land. Is this not getting two mous' profit for the labor, capital, and land of only one?" This passage comes directly from *Shen-shih Nung-shu.* See Ch'en Heng-li, p. 240, and also p. 36.
eSeveral sources including *Hu t'san-shu,* p. 20, and the *Wu-ch'eng HC* (1881), 28:19b, which is written by the same author and is substantially the same in content, quote the same passage from the *Wu-hsing chang-ku chi,* pp. 38-39. This passage reads, "One mou of good land will produce approximately 80 *ko*, each 20 *chin* being 1 *ko.* It is calculated that the cost of hoeing and cultivating it for one year does not exceed approximately 2 *liang*, and its profit is double that." The frequency with which this passage appears in various sources should not be taken as a sign of its accuracy.
fWu-ch'ing CC (1760), 2:7b. This passage also appears in *Wu-ch'ing CC* (1936), 7:17b, and *Nan-hsun CC* (1922), 30:25b-26b. It says that 1 mou of good land can produce 1,300-1,400 *chin.* It is calculated that the cost of hoeing and cultivating the ground for one year never exceeds 3 *liang;* thus the profit often doubles this.
gSilk, comp. Imperial Maritime Customs (Shanghai, 1881), p. 125.
hShuang-lin CC (1917), 14:3. This passage says that one mou of good land should yield approximately 1,000 *chin,* but those which exceed this figure are few, while those which do not reach it are many. In other words, this was nearly a maximum figure.
iShina shōbetsu zenshi, comp. Tōa dōbunkai (Tokyo, 1919-1920), XIII, 495.
j *A Survey of the Silk Industry of Central China,* comp. Shanghai International Testing House (Shanghai, 1925), pp. 28, 38, and 50. This survey reported on *Hu-sang* grown from seedlings in the Hangchow–Chia-hsing area, the *Hu-sang* in the Soochow area, and ungrafted *Lu-sang* in the Wusih area. These latter were spaced four feet apart, with about 300 trees per mou. They were only 3-4 feet high. Since the system of weights and measures was not standardized by the central government until after 1928, I have assumed that the 1925 figures of the Shanghai Testing House need to be converted to *shih-yung chih* figures. See also *Shina sanshigyō taikan,* comp. Sanshigyō dōgyōkumiai chūōkai (Tokyo, 1929), p. 82, for similar figures. See also Ch'en Heng-li, p. 39, where the author cites *Wu-hsing nung-ts'un ching-chi,* comp. Chung-kuo ching-chi t'ung-chi yen-chiu so (Shanghai, 1939), for similar figures. See also *Shina sanshigyō kenkyū,* comp. Tōa kenkyūjo (Osaka, 1943), pp. 82-83.
kD.K. Lieu, *The Silk Industry of China* (Shanghai, 1941), p. 7.
lYueh Ssu-ping, *Chung-kuo ts'an-ssu* (Shanghai, 1935), pp. 22, 70-71. See also *Chung-kuo shih-yeh chih: Chiang-su sheng,* ed. Chung-kuo shih-yeh pu, kuo-chi mao-i chü (Shanghai, 1933), V, 160, for Kiangsu figures.
mKōsoshō Mushakuken nōson jittai chōsa hōkokusho, comp. Minami Manshū tetsudō kabushiki kaisha (Shanghai, 1941), p. 44.
nCh'en Heng-li, p. 41. In revisiting the silk area of Wu-hsing and T'ung-hsiang in 1956, Ch'en found that, in general, yields of mulberry fields were much below the late Ming standard found in the *Pu Nung-shu.* Most places produced only 178-815 *chin* per mou. Only in one experimental station in Wu-hsing did he find the yield to be exceptionally high, 3,000 *chin* per mou. See pp. 41-43.

Ch'ing were difficult to emulate. In Wusih, for example, mulberry fields were more densely planted than in northern Chekiang, but they also had a lower yield.[26] This type of tradeoff was recognized earlier by the *Pu Nung-shu,* which had said, "In the method of planting, value sparsity."[27] Too many trees would crowd each other and compete for sun, air, and nutriment. But what the declining yields most clearly reflected was a lower standard of cultivation than had been practiced in Chia-hsing and Hu-chou in the early Ch'ing. Anything less than total devotion could drastically affect yields. One Ch'ing source said that if the soil was not cultivated four times a year, the yield would be cut in half. "Is this not getting two mous' profit for the land, labor, and capital of only one mou?" it asked.[28] In short, in mulberry growing, as in other areas of traditional Chinese agriculture, intensive cultivation, dependent on the patience and hard work of the peasant, brought optimal results.

SERICULTURE

The silkworm raised in China was the *Bombyx mori.* The type raised in central China produced two crops, or generations, a year, the main one in the spring and another in the summer. The spring crop molted four times, while the summer crop molted only three times. The silk produced by these worms was white in color and extremely fine in texture.[29] In south China, the silkworm was a "tropical" variety which could produce several generations each year. Its silk was softer and shinier, but coarser, than the central Chinese variety.[30] In contrast to the *Bombyx mori,* the wild silkworm of Manchuria and Shantung, which fed on oak leaves, was of a completely different family, the *Antheraea saturnidae,* of which *Antheraea pernyi* was the Chinese species.[31]

Traditionally it was considered inadvisable to raise a summer crop of silkworms, since the labor was needed for the planting of rice and other crops at that time. As the *Pu Nung-shu* put it, it would be a case of the "small obstructing the big." Summer worms should be raised, it said, only if there were extra mulberry leaves for which no other use could be found.[32] Even when summer worms were raised, their size and quality could not compare to those of the spring crop.[33] By the late nineteenth century, however, the market in Kiangnan for silk reeled from second-crop cocoons had expanded. In the old days, summer worms had been raised by women to earn a little extra pocket money but, since the expansion of foreign trade, this crop could produce a significant income, although never more than 30 to 40 percent of that generated by the first crop.[34] In the 1920s and 1930s, cocoon merchants began to promote a third, autumn crop of worms, but this practice never became widespread.[35]

The process of raising the main crop of silkworms required more than a month and a half of intense activity each spring. It took about eight days to hatch the silkworm eggs, about twenty-eight or twenty-nine days for feeding the worms, five days for the worm to spin its cocoon, about eleven days in the cocoon stage, and three days for the moth to lay its eggs.[36] During the month required for the feeding, the worms required constant attention. All other activities in the peasant household and in the village were suspended. People shut their doors and did not visit each other. The local yamen did not collect taxes or hold judicial proceedings. And all local schools closed so that the children could help out.[37]

According to the descriptions in sericultural manuals, at the time of *ch'ing-ming,* in early April, the silkworm egg-sheets, which had been carefully stored in a cool place over the winter, were taken out, and if the leaves on the mulberry trees were already "the size of a copper cash," then the process of hatching the eggs could begin. The egg-sheets were warmed inside a person's clothing, or placed under a quilt at night.[38] After the eggs hatched, the silkworms would be arranged evenly in shallow baskets in a warm room. They would then go through four molting stages, known in Chinese as four "sleeps." After each sleep, they would be fed mulberry leaves. During the first two stages, the leaves were chopped up rather fine and fed to the worms five or six times a day for about four days each. After the third molting, the pace quickened as the worms began to eat much faster. In the daytime, they had to be fed six or seven times and at dusk and during the night, once each. During the third stage, the leaves had to be cut up coarsely but, in the last period, they did not need to be cut up at all. Even small branches could now be given the worms. In the last stage, they had to be fed at least ten times a day, and the sound of their eating created a din.[39] The baskets had to be changed often so that the worms could be kept clean of their own droppings. The bigger the worms grew, the more often the baskets had to be changed and the greater the number of them needed.[40]

After the fourth molting, the silkworm was ready to spin its cocoon. At this stage the timing was crucial. If the worm was selected prematurely for spinning, it would not have eaten enough to produce the right amount of silk (see Figure 2). If, however, the worm was selected too late, it would have already spit out some strands of silk and would not form a plump cocoon.[41] In any case, the silkworms were placed on straw stacks —sometimes arranged on bamboo screens—upon which they were to spin their cocoons. This was known as "climbing the mountain."[42] During this process it was important to keep the room warm by building fires on all sides; otherwise, the worms would not spin their cocoons properly.[43] According to the well-known seventeenth-century technological manual, *T'ien-kung k'ai-wu*, using a fire was unique to Chia-hsing and Hu-chou and accounted for the durability of Kiangnan silk.[44]

After about five days, when the cocoons were completely formed, they were removed from the stacks and spread out in baskets in a cool room to await reeling. However, prior to this, a few of the best cocoons were selected for breeding purposes. The moths were permitted to break through the cocoons, mate, and lay eggs. These were placed on a sheet, which would be hung up to dry for a few days, until their color turned black.

FIGURE 2 Sorting the Worms

Then some lime powder would be sprinkled over them, and the sheets would be folded up and stored until winter. During the twelfth lunar month, the sheets were taken out and rinsed in salt water, or cold tea, for about ten days in order to kill off the weaker eggs. They would then be stored again until spring, when they would be taken out to hatch.[45]

In some parts of Kiangnan, it was customary for the peasants to purchase egg-sheets from the market rather than preparing their own. Although the purists advised against doing this, because they felt that the families who prepared their own egg-sheets would have better control over the quality of the eggs, the market for silkworm eggs was quite large and lucrative.[46] During the Ch'ing, the principal center for egg-sheets was Yü-hang in Hangchow prefecture and, in the twentieth century, Shao-hsing became a second center. Together they produced about 45 percent of the total eggs needed by farmers in Kiang-nan during the 1920s.[47]

It is evident from the Ming and Ch'ing sericultural manuals that extreme precariousness attended every step in the sericultural process. As the manuals said, if the right balance between starvation and fullness, between dryness and dampness, and between cold and warmth was not maintained, the silkworms would get sick and die. It was acknowledged that failures were due in part to human deficiencies, such as laziness, and in part to natural factors such as wind, rain, fog, and dew.[48] Silkworms thrived in a warm and dry climate, but mulberry trees needed warm and humid weather and rain rather than sunshine. So in a certain sense the requirements of the two were somewhat in conflict, creating even more complex needs. Good weather, combined with favorable market conditions, could create a small windfall for the peasant, but bad weather and a slow market could spell disaster.[49]

Since the weather was beyond human control, it is not surprising that sericultural practices involved a large element of superstition. It was commonly said not only that silkworms liked warmth and hated cold, liked dryness and hated dampness, liked cleanliness and hated dirt, but also that they liked quiet and hated noise, liked good fortune and hated disaster.[50] Silkworms also hated the odor of frying fish and meat, tears and shouting, and women who had just given birth in the past month.[51] They also hated pregnant women, smoke, vinegar, wine, musk, oil, damp leaves, the pounding of mortar, and mourning.[52] When Japanese observers in the 1920s asked peasants in Nan-hsun what accounted for the success of their silkworm crops, they replied that it was a combination of *p'u-sa (bodhisattva)* and good luck.[53] In his well-known short story, "Spring Silkworms," Mao Tun, who was a native of a leading silk district in Chekiang, poignantly depicted the hopes and anxieties of the peasants about their silkworms, and detailed the superstitions and rituals accompanying each step in raising them.[54]

After the weather, the second greatest threat to a silkworm crop was disease. Unfortunately for the peasant household, silkworm disease was not usually detectable until the final stage, just before the spinning of the cocoon, and there was no way to prevent the loss of the entire investment of leaves, the largest expense in raising worms. The extreme susceptibility of silkworms to pebrine disease was the most serious problem in sericulture. It was responsible for the European silkworm blight in the mid-nineteenth century. It is not clear when the disease spread to China, but by the early twentieth century it was rampant. By the 1920s, it was estimated that the silkworm egg-sheets sold on the market were 75 to 95 percent diseased.[55] In Japan and France, one ounce of eggs would yield 110–133 pounds of cocoons, whereas in China one ounce would yield only 15–25 pounds.[56] The failure to check this disease was the most critical technological factor in the decline of the Chinese silk industry in the twentieth century.

The French scientist Louis Pasteur, who had isolated and identified pebrine disease, invented a simple process for controlling it by means of microscopic inspection of silkworm moths and their eggs. For this technique to be effective, there had to be some central authority who could control the manufacture and distribution of egg-sheets. In Japan, for example, from the late nineteenth century only eggs produced by licensed farmers could be used in sericulture; individual households were forbidden by law to breed their own eggs.[57] In China, experimental egg-stations were not established until the 1920s at Hangchow, Chinkiang, and a few other localities, and central control was never established.[58] The egg-stations, however, were highly successful—although too late to halt the declining fortunes of the Chinese industry. They prepared egg-sheets which were scientifically inspected and distributed to peasants in the local areas. These "improved" eggs were clearly superior to the so-called "native" eggs; not only did the silkworms hatched from these eggs have a greater chance of survival, but the yield of silk was much higher. In Kiangnan, according to one scientific investigation, it took an average of 321 catties of dried cocoons produced from the improved eggs to produce 100 catties of raw silk, whereas it took an average of 486 catties of cocoons from native eggs to produce the same amount of raw silk.[59] It also appears that the silkworms from the former required fewer leaves than those from uninspected eggs. As Table 2 shows, the yields from leaves of improved silkworms matched, but did not exceed, recorded yields from the early Ch'ing.

SILK REELING

Those cocoons that survived were carefully sorted; only small and firm ones were suitable for reeling. Those which were too loosely formed, double-shaped, infected with maggots, or in any other way deformed, could not be reeled into raw silk.[60] Defective cocoons could, however, be used to make silk wadding as well as to weave rough silks. They were placed in a pot and boiled for a long time, after which they were stretched on a bow and dried in the sun.[61] The manufacture of this type

TABLE 2 Yields: Leaves to Cocoons

Date	Place (if specified)	Units of Leaves Needed to Yield One Unit of Cocoons
17th century	Chia-hsing	15[a]
19th century	Chia-hsing	10[b]
19th century	Kiangnan	20[c]
1920s	Kiangnan	15-23[d]
1920s	I-hsing, Shao-hsing	10[e]
1920s	Shih-men	17[f]
1920s	Kwangtung	16[g]
1920s	World, in general	25[h]
1930s	China, in general	20-25 ("native" eggs)[i]
1930s	China, in general	15-20 ("improved" eggs)[i]
1930s	Kiangnan	16[j]
1950s	Chia-hsing	14-15[k]

Notes: [a]Ch'en Heng-li, pp. 37, 47. Ch'en calculates that in the Ming it took 15.23 *shih-chin* of mulberry leaves to yield 1 *shih-chin* of cocoons. 1 *chin* = 1.19 *shih-chin*. See pp. 36-37, 293-300.

[b]*Chia-hsing FC* (1879), 32:22b. This passage says that it took 100 catties of leaves to produce 10 catties of cocoons.

[c]*Ts'an-sang chien-ming chi-shuo*, Huang Shih-pen (1882, 1888), 19b. This passage says that it took 2,000 catties of leaves to yield 150-160 *liang* of raw silk. Since, according to this source, it took 10 parts cocoons to produce 1 part raw silk, we can assume that it took 100 catties of cocoons to produce 10 catties of silk (i.e., 160 *liang*). Therefore it took 20 catties of leaves to produce every catty of cocoons.

[d]*A Survey of the Silk Industry of Central China*, pp. 7-8, 89-92. On p. 92, it says that 70 piculs of leaves were required to produce one picul of dry cocoons. Since on p. 14, it states that 3 parts of fresh cocoons were required to produce 1 part of dry cocoons, it took 23 parts leaves for every 1 part fresh cocoons. On p. 89, however, it says that only 14.8 to 16.8 piculs of leaves were required to produce 1 picul of fresh cocoons. Compare also pp. 7-8.

[e]*Shina sanshigyō taikan*, pp. 153-155.

[f]Ibid. The passage states that it took 200 catties of leaves to produce 12 catties of cocoons.

[g]"The Silk Industry in Kwangtung Province," *Chinese Economic Journal* 5.1:608 (July 1929), states that it took about 50 piculs of leaves to produce 1 picul of dry cocoons in Kwangtung.

[h]Leo Duran, *Raw Silk: A Practical Handbook for the Buyer* (New York, 1921), p. 17. Generally, 25 lbs. of leaves were required to produce 1 lb. of cocoons.

[i]Yueh Ssu-ping, p. 27. According to this source it took 4,000-5,000 catties of leaves to produce 200 catties of cocoons from "native"eggs and 300-400 catties of leaves to produce 20 catties of cocoons from "improved" (scientifically prepared) eggs.

[j]D. K. Lieu, p. xi. This material seems to be taken from Tseng T'ung-ch'un, *Chung-kuo ssu-yeh* (Shanghai, 1933), pp. 40-42. It states that 2,400 catties of leaves were required to produce 150 catties of cocoons.

[k]Ch'en Heng-li, p. 47. This source says that, in the 1950s, it took 14-15 *chin* to yield 1 *chin* of cocoons.

of wadding, and the rough silks sometimes woven from it, was a specialty of Chekiang, particularly Hu-chou. Peasants from Nanking and other localities in Kiangsu often sold their waste cocoons to be processed there. Within Hu-chou prefecture, Wu-ch'eng and Wu-k'ang were reputed to produce the best wadding.[62] An important by-product of silk manufacturing, wadding was used in quilted jackets and bedcovers and was much warmer than cotton padding. In Hu-chou, it was said that, while the local people seldom had the opportunity to wear the fine silks they produced, all could afford to wear clothes of silk wadding to ward off the cold.[63]

Those cocoons selected for reeling were stored in a cool room to delay the process of maturation as long as possible. Since the moth would break through the cocoon in about ten days, the reeling had to be completed within that period of time. The pressure during this period was intense, and the results sometimes showed the effects of haste.[64] Methods of killing the chrysalis by steaming, sunning, or soaking in a salt solution were recommended in Yuan and Ming dynasty manuals, but apparently they were not widely employed in the Kiangnan area during the Ch'ing period.[65] If they had been, the whole time pattern for silk reeling could have been spread out.[66] Hsu Kuang-ch'i, in his famous seventeenth-century work, *Nung-cheng ch'üan-shu,* explained that these methods had been used in north China, where there was not always enough labor to reel off the whole crop of cocoons immediately; in south China there was no need to kill the chrysalis since there was no labor shortage.[67] This step was to be avoided if possible, because the heat damaged the luster of the silk reeled from the cocoon.[68] F. Kleinwachter, a Maritime Customs inspector who had a special interest in the silk industry, reported from Chinkiang in 1879 that, "unless sufficient hands can be employed to finish reeling in time, the cocoons are steamed, to kill the chrysalis before reeling; but the silk from the steamed cocoons is not so good and shining."[69] Other foreign observers, however, felt that the practice of reeling silk hastily from live cocoons was a greater cause of the poor quality of Chinese silk.[70]

In any case, when the cocoons were ready to be reeled, they were placed in a basin of water at the foot of the reeling machine. Usually the water was heated by means of a small fire, although in earlier times cold water had also been used.[71] The cleanliness of the water was considered to be of paramount importance; the quality of Hu-chou silk, for example, was often attributed to the water in that area.[72] The hot water would loosen the ends of the silk, which would be picked out by the operator and threaded through eyelets, over and around various rolls, and finally onto the reel itself.[73] In most places during the Ch'ing, the reel was turned by means of a treadle although, until the 1870s, the hand-reel was still used in the Canton delta.[74] (See Figure 3, which shows a hand-reel.) In the smaller treadle machines, the operator was seated at the front, close to the basin of water. Occasionally two operators would be seated together. In the larger machines, one person might oversee the basin of cocoons, while a second stood at the back operating the treadle.[75] Usually, at the rear of the machine one or two small charcoal fires were built to dry the silk filaments as they were reeled, so that they would not stick together. The larger machines were usually operated by men and, since they required a greater investment, were probably used only in localities highly specialized in silk production.

There were many varieties of raw silk reeled in Kiangnan, but the basic distinction was between fine and coarse silk. Fine silk was used for weaving satins and for making warp, while coarse silk was used in ordinary grades of fabrics such as *ch'ou* or *ling*.[76] After the opening of the treaty ports, the fine silks were in greater demand abroad, while the coarse silks continued to be used for local weaving.[77] In the export trade, the fine silks were known as "Tsatlees," while the coarse silks were known as "Taysaams."[78] Chia-hsing and Shao-hsing were the main centers for Taysaams, while the Nan-hsun district of Hu-chou was the center for Tsatlees.[79]

As steam filatures were introduced in the Canton delta and the Shanghai area in the late nineteenth century, the reeling of silk was gradually separated from the sericultural process and

FIGURE 3 Reeling the Silk

was no longer a domestic enterprise in those localities affected by the filatures. Households would sell their fresh cocoons to local cocoon hongs or markets which were equipped with large ovens used to dry the cocoons.[80] The cocoon hongs would in turn sell the dried cocoons to the filatures. Usually it took three to three and one-half catties of fresh cocoons to yield one catty of dried cocoons. Typically the ratio of 3:1 was used in conversions.[81]

Filatures in Shanghai were usually two- or three-story brick buildings, which housed the offices of the firm as well as the factory. The factory area itself was divided into five sections: cocoon peeling, cocoon sorting, silk reeling, silk finishing and rereeling, and waste preparing. Only steam power was used in the filatures, and the same engine that supplied power also supplied the hot water for boiling the cocoons.[82] Most of the reeling equipment used in Shanghai was of "Italian" or Tavelle style, but in the Canton delta the "French" or Chambon type was used, even though it was considered inferior because only two ends could be reeled at one time. Later in the 1920s, some filatures in Hangchow and elsewhere advanced from the Tavelle to even more efficient Japanese machinery.[83]

Cocoons peeled and selected for reeling were then boiled and brushed in a basin for about five minutes. In Shanghai, girls from eight to twelve years old were given the unpleasant and steamy task of tending these basins and finding the ends of silk loosened from the cocoons.[84] The actual job of attaching the loose ends to the mechanically rotated reels was entrusted to older women. The thicker the silk thread desired, the greater the number of cocoons needed. After the silk was reeled, it was rereeled in order to make it more even. Finally it was packed in skeins to be shipped.

Whatever the advantages of mechanized reeling, labor-saving was not necessarily among them. It appears that the average factory worker could produce no more silk in one day than an average domestic reeler. Most late Ming and Ch'ing sericultural manuals state that one reeler could produce twenty to thirty *liang* of raw silk in one day, depending on the thickness of silk thread.[85] Twentieth-century sources generally give a lower rate for domestic reeling, ranging from twelve to sixteen *liang* per day.[86] Still, it appears that a factory worker could reel only eleven or twelve *liang* at best, and might reel as little as eight *liang* a day.[87]

Nor did mechanized reeling enable better yields of silk to be obtained from cocoons. Practically all sources since the seventeenth century are unanimous in stating that it took between eight to thirteen catties of fresh cocoons to yield one catty of raw silk through domestic reeling. As Table 3 shows, however, with mechanized reeling and the use of dried cocoons, there was, if anything, a decline in yields. These figures also reflect the deterioration in the quality of cocoons in the twentieth century, due not only to disease but, in the opinion of some observers, to increasing carelessness on the part of peasants when they were not planning to reel the cocoons at home.[88] Using cocoons raised from "improved" eggs, the filatures obtained higher yields, as good as 12:1, but this was no better than yields obtained on traditional domestic reeling machines. A yield of 18:1, or 15:1, was considered by experts to be more typical for the twentieth century.[89] The great advantage of steam filature reeling lay neither in labor-saving nor in cost-saving, for it did neither. Its unassailable advantage was that it produced raw silk of absolute uniformity and reliability. This was made possible not only by modern machinery, but also by factory production, which imposed regulations and standards that could not be achieved in domestic production. This standardization was demanded by foreign buyers of raw silk and, for it, they were willing to pay significantly higher prices (see Table 14, below).

PROCESSING AND WEAVING

By the twentieth century, in order to compete with steam filature silk in the export market, it was customary for domestically reeled silk in Kiangnan to be rereeled to remove imperfections and to produce a more standardized product. Like reeling, rereeling was usually a domestic enterprise, and it was centered in the Hu-chou area.[90] Sometimes, however, the silk hongs, or *ssu-hang,* would set up workshops and hire workers to do the job.[91]

Raw silk that was not exported had to go through a few intermediate processes before it could be woven into cloth. With traditional technology, the raw silk was first wound by hand onto spools,[92] and then made into warp and weft suitable for weaving. The warp (*ching*) generally consumed four parts of silk out of every ten, while the weft (*wei*) took six out of every ten. The warp was made by drawing the silk fibers through a large and elaborate frame, while the weft was made by spinning the fibers on a wheel very similar to a cotton spinning wheel.[93] In Europe, where these steps were done mechanically, this was called "throwing" the silk. Its basic purpose was to twist several fibers into one strand strong enough for weaving. In China, if the warp was prepared in the peasant household from local material it was called *hsiang-ching* (lit., "country warp"), and it was sold to the *ching-hang,* or warp hongs. If the warp was prepared from raw silk received on consignment from a *ssu-hang,* or silk hong, it was called *liao-ching* (lit., "material warp").[94] Another important process was the dyeing of the silk. Although for simple domestic weaving this could be done at home, silk prepared for weaving in urban workshops was dyed on consignment from the silk hongs. Traditionally, the dyeing almost invariably took place before the weaving. After modern weaving factories were established in the twentieth century, the silk was often dyed after it had been woven.[95]

In the weaving of fine and elaborate silks, traditional Chinese artisanship had no peer. The varieties of high-grade fancy silks and satins were innumerable. These were woven in urban workshops on large and complex wooden draw-looms, which

TABLE 3 Yields: Cocoons to Raw Silk

| Date | Place (if specified) | Units Fresh Cocoons Needed To Yield One Unit Raw Silk[a] | |
		Domestically reeled	Machine-reeled
12th century	An-chi	15.5[b]	—
17th century	Chia-hsing	12.5[c]	—
19th century	Kiangnan	10 (8–13)[d]	—
19th century	Kiangnan	11[e]	—
1910s	Shuang-lin	8[f]	—
1920s	Kiangnan	—	10.5–21[g]
1920s	Kiangnan	—	13.5–16.5[h]
1920s	Kwangtung	15[i]	
1920s	World	12[j]	—
1930s	Wu-hsing	—	18[k]
1930s	Kiangnan	10–13[l]	18[m] ("native" eggs)
1930s	Kiangnan	—	12[m] ("improved" eggs)
1950s	Chia-hsing	—	23[n]

Notes: [a]These figures are based on the formula 3 parts fresh cocoons were equivalent to 1 part dry cocoons. See *A Survey of the Silk Industry of Central China,* p. 14, and D. K. Lieu, pp. xi, 136. *Shina sanshigyō kenkyū* p. 108, states that it took 3.2–3.8 parts fresh cocoons to yield 1 part dry cocoons, and Yueh Ssu-ping, p. 180, gives 3.2 as an average. Taxes were calculated using a 3:1 ratio, according to *Shina sanshigyō taikan,* p. 453.

[b]*Nung-shu,* Ch'en Fu (1154; *Chih pu-tsu chai ts'ung-shu* ed.), 3:4a, says each catty of cocoons would yield 1 *liang* and 3 *fen* of silk, or 1.03 *liang.*

[c]Ch'en Heng-li, pp. 48–49. 12.5 catties of fresh cocoons were needed to produce 1 catty of raw silk.

[d]*Ts'an-sang chien-ming chi-shuo,* 33a. *Kuang ts'an-sang shuo chi-pu,* Chung Hsuehlu (1862, 1877), 2:26. And also *Chia-hsing FC* (1879), 32:22 a–b. These sources all state that it took 100 catties of cocoons to yield 10 catties of raw silk. But *Ts'an-sang chien-ming chi-shuo,* 1b, and *Kuang ts'an-sang shuo chi-pu,* 2:26, also state that 1 catty of cocoons would produce from 1.2, or 1.3, to 2 *liang* of raw silk or, in other words, it took 8 to 13 catties of cocoons to yield 1 catty of raw silk.

[e]*Silk,* p. 53. This source states that 100 catties of cocoons would yield 9 catties of raw silk.

[f]*Shuang-lin CC* (1917), 14:2. This source states that 1 catty of cocoons would yield 2 *liang* of silk at best.

[g]*A Survey of the Silk Industry of Central China,* pp. 14–16, states that it took 3.5 to 7.0 catties dry cocoons to yield 1 catty of raw silk. On p. 68, it states that 1 catty of dry cocoons would yield 1.2 *liang* of silk. In other words, it reported about 12:1 yield of fresh cocoons to silk. On p. 92, however, the report states that 6 piculs of

dry cocoons would yield 1 picul of silk, or about an 18:1 ratio of fresh cocoons to silk; 15:1 is the standard generally preferred by them.

[h]A number of Japanese surveys state that it took 450–550 catties of dry cocoons to yield 100 catties of raw silk, or 13.5–16.5:1 ratio of fresh cocoons to raw silk. See for example, *Shina sanshigyō kenkyū*, p. 108; *Shina sanshigyō taikan*, p. 269; and *Shina shōbetsu zenshi*, XIII, 578.

[i]"The Silk Industry in Kwangtung Province," *Chinese Economic Journal*, p. 608, states that 1 catty of raw silk could be produced from 5 catties of dried cocoons.

[j]Duran, p. 17.

[k]D. K. Lieu, pp. 136, 138, states that it took 6 parts dry cocoons to yield 1 part of raw silk.

[l]*Chung-kuo chin-tai shou-kung-yeh shih tzu-liao* (Peking, 1957), ed. by P'eng Tse-i, III, 694–695. These data are collected from 24 localities in Chekiang.

[m]Yueh Ssu-ping, p. 27, states that 100 catties of dried cocoons raised from "native" eggs would yield 16–17 catties of raw silk, while 100 catties from "improved" eggs would yield 25 catties of raw silk.

[n]Ch'en Heng-li, p. 207, states that it took 7.6 catties of cocoons to yield 1 catty of raw silk reeled by machine. This undoubtedly means 7.6 catties of dry cocoons.

required two or three weavers, usually men, to operate, one of whom perched on top of the frame in order to control the raising of the warp threads to create a design (see Figure 4). The draw-loom was known in Chinese as the *hua-lou chi* (lit., "pattern-frame loom"). On the other hand, there were many smaller looms for weaving more ordinary types of silks, gauzes, ribbons, and so on. The simplest of these were the *yao-chi,* or waist-looms, which could be operated at home by one woman.[96] Silk weaving was an extremely specialized profession. Each district and each workshop had its own types and patterns of silk, which in some cases they had maintained for centuries.[97]

Although weaving remained predominantly a traditional handicraft, in the early twentieth century, a few modern silk weaving factories were established, the first in Hangchow in 1905.[98] These factories were equipped with iron Jacquard looms, similar to those already used in Japan.[99] Electric power looms did not follow until quite a bit later, the first one being introduced at Shanghai in 1915, and the second at Hangchow in 1921.[100] The establishment of these modern mills ushered in a revival of the weaving industry in the 1910s and 1920s.

FIGURE 4 Weaving Silk on a Draw-Loom

MODERNIZATION REDEFINED

While it was customary for foreign merchants in the late nineteenth century, echoed by their Chinese counterparts in the twentieth, to lament the technological backwardness of the Chinese silk industry, the real causes of the industry's problems were not simply technological. In many respects, the traditional Chinese technology was not backward at all, and in fact compared quite favorably to modern technology. For example, as just noted in Table 3, the yields of silk from cocoons did not necessarily improve with steam filature reeling. Yields in mulberry planting, as well as yields of cocoons from mulberry leaves, as seen in Tables 1 and 2, had both reached a high level by the late Ming or early Ch'ing—one which nineteenth and twentieth century peasants found difficult to match. Even the results obtained by Japanese sericulturalists in the 1920s, after decades of scientific experimentation, barely exceeded the best results achieved by the late Ming and Ch'ing Chinese experts.[101]

Traditional Chinese technology was well in advance of the Japanese until the nineteenth century. The treadle reeling machine, for example, was not introduced in Japan until the early Meiji period, when it was hailed as a great improvement over the hand-operated machine because it freed the operator's hands to deal with the cocoons, and also because it drove more power to the reel, which could thus be larger.[102] In China, the treadle reeling machine was probably in use by the Sung period, if not earlier. Two extant Yuan manuals, Wang Chen's *Nung-shu* and the *Nung-sang chi-yao,* both explicitly identify and discuss the treadle machine. Certainly it was in widespread, although not universal, use during the Ming and Ch'ing.[103]

If traditional technology was so effective centuries before the opening of the treaty ports, two related questions must be asked: Why did this technology not continue to improve significantly after the Ming period? and Why did it fail to improve even with the stimulus of foreign markets after 1842? Although the traditional techniques were impressive, the technology of sericulture and silk manufacturing seems to have leveled off, and even stagnated, after the Ming or early Ch'ing.[104] And the opening of the treaty ports did nothing to reverse this trend. By contrast, the Japanese silk industry, which had barely matched the best Chinese standards by the early twentieth century, soon went on to outstrip them. Between 1900 and 1950, the Japanese were able to increase 3.7 times their overall yield of raw silk from one *tan* of mulberry field and more than double their yield of raw silk from cocoons.[105] By 1924, about 94 percent of Japan's raw silk output was machine-reeled.[106]

In short, it could be said that the Japanese silk industry had "modernized," while the Chinese had not. But modernization did not simply mean the improvement of yields of silk per units of land, labor, or capital, or the application of inanimate sources of energy to the reeling process. Modernization in Japan meant the reduction of those risks, natural and commercial, which were so characteristic of silk manufacture. It meant producing not just better silk, but *consistently* better silk. Control of silkworm egg disease was a major step in that direction. Mechanized reeling was the second critical aspect of standardization. Neither required a very complex technology. What they did require was centralized leadership, authority, and organization. The scarcity of these critical factors in China suggests that the problems of the Chinese silk industry in modern times were economic and institutional in nature, not technological.[107]

TWO

The State and Traditional Enterprise

In China, the manufacture of silk had developed not only an advanced traditional technology, but also an advanced form of traditional economic organization. Although it was theoretically possible for a peasant household to engage in all the stages of silk manufacture from the growing of mulberry trees to the weaving of fabric, in reality no household was completely self-sufficient. By Sung times, there were already markets for silkworm eggs, mulberry leaves, and reeling and weaving equipment, and a wide range of local and interregional merchants facilitated this trade.[1] Although sericulture and reeling remained basically domestic activities, the weaving of silks, particularly those of high quality, tended to become a specialized enterprise. Weaving required more complex technology than the other stages; furthermore, the large looms for luxury silks required a greater investment. The household production of raw silk and simple fabrics lay at one end of the spectrum of organizational possibilities, the enterprise of weaving luxury fabrics at the other.

Besides relative complexity, the manufacture of silk had another characteristic: since ancient times it had been closely sponsored and controlled by the state. Not only was silk part of the tax system but, ever since the Han dynasty, the Imperial court had established factories for the weaving of silk fabrics for official use. The court used these silks for ceremonial robes and ordinary clothing, as a means of payment to civilian and military officials, and as a vital commodity in trade and tribute relations. Not until the Sung period was the silk industry commercialized to any significant degree,[2] and even in the Ming and Ch'ing the government still constituted the single largest source of demand for silk.

In the 1950s and 1960s, the development of silk manufacture during the late Ming and Ch'ing attracted the attention of several Chinese and Japanese scholars who were interested in the question of the "sprouts of capitalism." In mulberry cultivation, these scholars stressed that leaves were produced not just for the peasants' own use, but also for an expanding market. They also pointed to significant improvements in silk technology, which were stimulated by a growing urban market for silk and silk products. But it was the weaving industry that aroused the keenest interest of these historians, who maintained, first, that the Imperial Silk Weaving Factories (Chih-tsao chü), which embodied the old feudal order, were gradually replaced by private enterprise. Second, they emphasized that, with the breakdown of the Ming compulsory artisan service for the Imperial Factories, weavers could be freely hired on the open market and even developed a class consciousness. Third, they argued that private weaving was organized in a putting-out system and eventually developed into "manufacture," or factory production. These trends constituted the "sprouts of capitalism" in the silk weaving industry.[3]

The concept of "the sprouts of capitalism" was first popularized in the 1950s by Shang Yueh, who claimed that China had exhibited in the late Ming the signs of indigenous growth toward capitalism—commercialization of agriculture, monetization of the economy, development of handicrafts, and the rise of new urban centers. Only the Manchu conquest of China and the invasion of Western imperialism destroyed these sprouts and prevented them from maturing into capitalism. While Shang Yueh's views satisfied nationalistic impulses, they were later attacked by critics for underestimating the destructive role of Western imperialism in turning China into a semi-feudal and semi-colonial country, and for implicitly undermining the historic anti-imperialist role of the Chinese Communist Party.[4] A considerable body of writings on this academic controversy emerged in both China and Japan during the 1960s.[5]

This "sprouts" controversy commands our attention, not as a political polemic, although it certainly has that aspect, but because it touches on an issue vitally important to all students of Chinese history: namely, that of the nature of traditional Chinese economic organization and its capacity for change caused by either endogenous or exogenous factors. While it is my view that no real "sprouts" ever existed, a systematic review of the arguments and evidence concerning the development of the silk weaving industry reveals that important changes did take place and helps us to understand them in historical perspective.

IMPERIAL SILK WEAVING FACTORIES AND THE KIANGNAN SILK MARKET

During the late Ming and the Ch'ing, three great Imperial Silk Weaving Factories manufactured virtually all of the high-quality silks for the Imperial court. These three factories, at Nanking, Soochow, and Hangchow, had been established in the late tenth century, functioned at least partially during the Yuan, and were revived under the Ming.[6] During the early Ming, the Kiangnan factories were only three members of a system of about twenty-five Imperial Factories, of which those located at the capitals of Peking and Nanking were the most important. Of the others, thirteen were located in Chekiang and Kiangsu provinces, and the rest in Anhwei, Fukien, Kiangsi, Szechwan, Honan, Shantung, and Shansi provinces.[7] However, by the Ch'eng-hua reign (1465-1487), many of these factories were being phased out, and it became the custom for those localities in which sericulture was less developed to meet their quotas for tax payments by purchasing silk goods from the major weaving centers. By the late Ming, only a handful of the local factories still functioned, leaving Soochow and Hangchow as the main centers.[8]

During the Ch'ing, initially there were only four Imperial Silk Weaving Factories, one at the capital of Peking, and the three in Kiangnan, at Soochow, Hangchow, and Nanking but, by the eighteenth century, the Peking factory had ceased to be of any importance. It is thought that most of the silks for the use of the court itself were woven at Peking or Nanking, while the Soochow and Hangchow factories specialized in weaving tribute silks.[9] A very limited quantity of silk was woven by the factories for private use.[10]

Because of the complexities of the imperial system for procuring silks, and because of the miscellaneous nature of the historical records available, it is difficult to derive a comprehensive picture of the total demand for silk generated by the Ming and Ch'ing imperial governments.[11] There is no question, however, that a very large quantity of silk was used by the Ming court.[12] During the Wan-li period (1573-1619), the official annual quota for all the factories was only about 47,000 bolts, of which 31.9 percent was manufactured at Peking, 7.2 percent at Nanking, and 60.9 percent at the local factories.[13] However, these figures represented only the official quota and did not include the special orders which were given with increasing frequency to the local factories during the Ming. According to the *Ming-shih*, from the middle of the Wan-li period, Soochow and Hangchow alone were manufacturing 150,000 bolts of silk annually.[14] In the peak year of 1604, the five prefectures of Soochow, Sungkiang, Hangchow, Hu-chou, and Chia-hsing produced 180,000 bolts for the court.[15]

During the early Ch'ing, it appears that the output of the three Kiangnan factories increased substantially. In the mid-sixteenth century, for example, the Nanking Imperial Factory had only 300 looms, while Soochow had 173 looms.[16] These figures are small compared with the respective figures of 538 and 800 for the early Ch'ing, as shown in Table 4. The productive capacity of all three factories seems to have reached a peak during the K'ang-hsi period and then to have declined. This can be seen from a reduction in the expenses of the factories from 452,300 taels in the early K'ang-hsi period[17] to approximately 345,000 taels around 1708.[18] Between 1725 and 1812, the amount of money spent for artisans' wages for the three factories was also reduced by 35 percent, from 213,443 taels in 1725, to 140,055 in 1812.[19]

One reason why it is difficult to get a clear picture of the output of these factories is that their financing was, as one historian has said, "chaotic."[20] Although in 1664 it had been decided that the expenses of the Imperial Factories should be jointly underwritten by the Board of Works and the Board of Revenue,[21] in reality funds came from dozens of sources, including the provincial treasury, shipbuilding funds, salt revenues, and elsewhere. The management of the Imperial Factories was politically sensitive and extremely lucrative. During the late Ming, the factories had fallen under the control of court eunuchs, whose authority had been deeply resented by local officials and workers.[22] During the Ch'ing, the operation of the Imperial Factories was entrusted to Manchu bondservants, who were outside of the regular bureaucracy and responsible personally to the emperor. In his study, *Ts'ao Yin and the K'ang-hsi Emperor,* Jonathan Spence has vividly shown how the Ts'ao family virtually monopolized the posts of Imperial Textile Commissioners (Chih-tsao) during most of the K'ang-hsi and Yung-cheng reigns (1662–1735), from which they

TABLE 4 Number of Looms in the Imperial Factories
During the Ch'ing

Date	Nanking	Soochow	Hangchow	Total
Early Ch'ing[a]	538	800	770	2,108
1725[a]	557	710	750	2,017
1745[a]	600	663	600	1,863
1870s[b]	545	245	115	905

Notes: [a]P'eng Tse-i, "Ch'ing-tai ch'ien-ch'i Chiang-nan chih-tsao ti yen-chiu," *Li-shih yen-chiu*, 1963.4:99–100. Ts'ao Yin, the Imperial Textile Commissioner at Nanking from 1692 to 1712, reported that there were a total of 664 looms at the Nanking factory and 2,500 artisans. See Jonathan D. Spence, *Ts'ao Yin and the K'ang-hsi Emperor: Bondservant and Master* (New Haven, 1966), pp. 89–90.

[b]*Silk*, pp. 62, 113. Compare Shih Min-hsiung, *Ch'ing-tai ssu-chih kung-yeh ti fa-chan* (Taipei, 1968), p. 24.

exercised enormous influence over the bureaucracy of the Kiangnan region.[23]

During the Ming, the raw silk for the factories was provided through the tax system. The silk-growing localities paid a summer tax in raw silk instead of grain, which was effectively designated for the use of the factory nearest to it.[24] It is important to note that the Ming government's share of the silk market far exceeded the quantity of silk woven at the Imperial Factories. In addition to raw silk, the silk districts were assessed taxes in *chüan*, an ordinary silk fabric woven domestically. This was used by the government to pay officials as part of their salaries.[25] Table 5 gives a rough idea of the magnitude of the government's share of the market in silk fabrics beyond the output of the Imperial Factories. It should be noted that, while the Ming government in the Wan-li period collected as many as 200,000 bolts of *chüan* per year as tax silk and authorized the weaving of as many bolts of higher-grade silks, the official quota for all the Imperial Factories together was only about 47,000 bolts of satins and other luxury silks, as mentioned above.

During the Ch'ing, silk no longer played such a large role in the tax system. The silk districts in Kiangnan were still obliged

TABLE 5 Official Silk (*Kuan-chüan*) Collected in the Ming (in bolts)

Year	China	Chekiang province
1393	288,847[a]	139,140[a]
1485	280,000[b]	
1505–1520 annual average	126,767[c]	
1522–1562 annual average	320,459[c]	
1567–1571 annual average	288,358[c]	
ca. 1593		97,365[d]
1602	148,129[c]	
1620–1626 annual average	206,282[c]	

Notes: [a]Saeki Yūichi, "Mindai shōekisei no hōkai to toshi kinuorimonogyō ryūtsūshijō no tenkai," *Tōyō bunka kenkyūjo kiyō* 10:397 (November 1956). The exact source for the Chekiang figure is not cited, but apparently it is the *Ming hui-tien*. Saeki gives the figure 39,140, apparently a typographical error. Compare Mi Chu Wiens, "Socioeconomic Change during the Ming Dynasty in the Kiangnan Area," (PhD dissertation, Harvard University, 1973), p. 131, n. 88.

[b]Saeki Yūichi, "Min zenpanki no kiko," *Tōyō bunka kenkyūjo kiyō* 8:169 and 195, n. 11 (March 1956). The source is the Ch'eng-hua edition of the *Shih-lu*.

[c]P'eng Tse-i, "Ts'ung Ming-tai kuan-ying chih-tsao ti ching-ying fang-shih k'an Chiang-nan ssu-chih-yeh sheng-ch'an ti hsing-chih," *Li-shih yen-chiu* 1963.2:54, n. 5 (February 1963). P'eng's sources were the *Shih-lu* and also Liang Fang-chung, "Ming-tai hu-k'ou t'ien-ti chi t'ien-fu t'ung-chi," in *Chung-kuo chin-tai ching-chi shih yen-chiu ch'i-k'an* 3.1:95 (May 1935).

[d]P'eng Tse-i, ibid., p. 46. Original source not clear.

to pay tribute silk, *kung-ssu,* to the central government, but the amounts were very small compared to the Ming tax in silk. For example, in 1398, the nine hsien of Hangchow prefecture paid summer tax in silk amounting to 25,590 catties[26] but, in the late Ch'ing, the two prefectures of Chia-hsing and Hu-chou paid only 4,000 catties each of tribute silk before the Taiping Rebellion, and only about 2,000 catties afterwards.[27]

Instead, the Imperial Factories purchased their raw silk on the open market. Although they were supposed to pay the market price for raw silk, in fact they were able to exert considerable influence over the price structure.[28] The opportunities for corruption and profiteering by officials, both high and low, were great.[29] Nevertheless, although the state continued to play a leading role in the raw silk market in Kiangnan until the nineteenth century, its consumption of raw silk decreased from the middle of the Ch'ing period. In part this may have reflected the declining role of silk in state finances, as money played a larger role in official transactions, including tributary ceremonies. Another factor was that, by the nineteenth century, the Imperial Household Commission (Nei-wu fu) and the Board of Revenue (Hu pu) had built up a considerable inventory of silk, which was "sufficient for a hundred years."[30] This helps to account for the relative inactivity of the Imperial Factories during the nineteenth century.[31]

With the expansion of the commercial market for silk during the Ming and Ch'ing, the government's share of the total market declined as well. In the same period that government weaving of silk was being concentrated in the three main Kiangnan factories, the activity of silk weaving was spreading from the major cities to a number of marketing towns, or *chen,* in the T'ai-hu region. Saeki Yūichi observes that this happened when urban weavers, fleeing from their official obligations in the cities, brought with them advanced types of looms which permitted the weaving of better silks.[32] Among the best known of these *chen* were Sheng-tse and Chen-tse in the Soochow area, P'u-yuan and Wang-chiang-ching in Chia-hsing, and Shuang-lin, Wu-chen, and Nan-hsun in Hu-chou.[33]

One often quoted passage describes the growth of Sheng-tse as a weaving center:

> Previously in the Sung and Yuan, the weaving of *ling* and *ch'ou* was undertaken only by people of the prefectural city [Soochow]. Not until the Ming [Hung-]hsi and Hsuan-[te] reigns (1425–1435) did the local people gradually begin to weave silk, and still they often hired people from the prefectural city to do the patterns. After the Ch'eng-hua period (1465–1487), the local people started to excel in this profession, and it became widespread. Thus, within forty or fifty *li* of Sheng-tse and Huang-hsi, the people all pursued the profits of *ling* and *ch'ou.* Those who were wealthy hired artisans to make the patterns, while the poor did their own weaving and ordered their children to do the pattern work [i.e., operate the draw-loom]. As for the women, if they were not engaged in spinning, they

would be reeling silk day and night. Boys and girls from the age of ten years all struggled from morning to night to eke out a living. The supply of raw silk and the price of *ling* and *ch'ou* could spell the difference between surviving or not surviving for the ordinary people.[34]

Although the growth of these weaving centers occurred outside the government's direct control, the continued presence of the Imperial Factories and the high level of government demand for silk in the Ming undoubtedly provided the context in which this development could take place. By the Ch'ing period, however, other trends developed to provide a challenge to the preeminence of the state: a flourishing urban market, greater interregional trade,[35] and greater overseas trade, both legal and pirate. The fact that Ch'ing officials in the eighteenth century placed strict restrictions on the amount of first-quality raw silk that could be exported, fearing that foreign trade had caused the price of raw silk to soar, is one clear indication of the eroding influence of the government's demand.[36]

Thus, with respect to the silk weaving industry of Kiangnan during the Ming and Ch'ing, we can observe some long-term developments. The first was that the weaving of high-quality silks for the court, such as *tuan*, became concentrated in three Imperial Factories instead of being distributed among several. Second, the weaving of other types of luxury silks, such as *ling* and *ch'ou,* spread to medium-size urban centers, while the weaving of ordinary silks such as *chüan* continued to be widely dispersed among many localities, both urban and rural.[37]

Third, although it is true, as the proponents of the "sprouts of capitalism" suggest, that the manufacture of silk for commercial purposes greatly expanded in the late Ming and Ch'ing, nevertheless, the government played a significant role in this process. Perhaps one might say that the expansion of government demand and the growth of the commercial market were interdependent. The expansion of the commercial sector had been to a large extent sparked by government demand, but at the same time the government was increasingly dependent on the commercial sector for fulfilling its needs. These needs were, however, declining over time, and were largely expressed indirectly through commissioned work rather than directly through government production. The recruitment of weavers by the Imperial Factories provides a clear example of this tendency.

WEAVERS FOR THE IMPERIAL FACTORIES

From the Ming to the late Ch'ing, the organization of production in the Imperial Factories and the recruitment of weavers for them progressed through several stages. During the Ming, the Imperial Factories were manned by weavers who were officially registered as artisans in the government's *Yellow Registers* (*Huang-ts'e*), which were the basic census records maintained by the Board of Revenue for the purpose of taxation. Registered artisans were exempt from taxation and miscellaneous compulsory services (*yao-i*), but were required to serve at specified times in the various government workshops for weaving, porcelain manufacture, brick-making, and other trades. The essential characteristic of artisan status was that it was permanent and hereditary. The artisan could not, in theory, free himself or his children from his official status and its obligations.[38]

There were several categories of registered artisans. One was that of "resident" artisans, or *chu-tso chiang,* who resided at Peking or Nanking and served regularly at the factories there, forming the major part of their labor force.[39] Because of their resident status in the capitals, these artisans were closely tied to the Imperial court and were supervised by the eunuch officials of the Inner court.[40] The "rotating" artisans, or *lun-pan chiang,* constituted the second major category. They were not continuously in residence at the factories, but served there by rotation under the Board of Works. The normal pattern was a period of three months' service every three years at the capital, but there were variations allowed according to the geographical proximity of the artisan to the capital.[41] Although the general management of the Imperial Factories was under the supervision of eunuch officials, the internal management of each was handled by several *t'ang-chang* (lit., "hall heads"), who were selected from the wealthiest and most influential of the weavers to serve by rotation. They were responsible for procuring and distributing the raw silk, supervising the work of the weavers, and shipping the finished products to the capitals.[42]

During the fifteenth century, the system of registered artisans began to suffer abuses and irregularities, which eventually led to its demise. The artisans found their obligations onerous, and many escaped service simply by fleeing, or neglecting their duties.[43] Since traveling to and from the capital involved great inconvenience and expense, commutation of the weavers' obligations to silver payments was allowed in 1485 and, by 1562, it was made compulsory for all provinces.[44] A contrary tendency was that in which wealthy civilian households (*min-hu*) surrendered themselves to artisan status in order to gain exemption from taxes and miscellaneous services. Such registration was, of course, in name only; in fact the household could avoid actual service as weavers by payment of a monthly sum of cash to the local officials, who in turn would use the money to hire outside weavers.[45] Commutation of services and the decline of registration was part of the general tendency in the late Ming toward the simplification of taxes and obligations, and their payment in silver—which eventually resulted in the so-called Single-Whip reforms.

With the breakdown of the registered artisan system, the Imperial Factories sought alternative methods of meeting their quotas. Some of the smaller factories outside of the Kiangnan area, which had actually stopped their own operations by the mid-Ming, simply purchased the necessary silk fabrics from markets at the major centers.[46] Other factories—particularly in localities where the specialized skills for the weaving of high quality silks were lacking—were increasingly forced to recruit private weavers, that is, weavers not of official artisan status.[47] Another way by which the factories filled their quotas was to place orders with private weavers on the outside through a contractor who was known as a *pao-lan jen*. This practice was called *ling-chih* (lit., "receive weaving"). The contractor would distribute the raw silk, or the money to buy it, to the private weavers. The latter in turn would weave the silk and return it to the contractor. Like the *t'ang-chang*, the contractor's duty was to serve as a guarantor. Unlike the *t'ang-chang*, however, the contractor worked outside the factory. In Soochow and Hangchow, this system was already well established by the first half of the fifteenth century; and by the sixteenth century it had already become legal.[48]

Initially, the Ch'ing authorities continued the basic practice of *ling-chih*. Rich households in the local area, not necessarily weavers themselves, were selected to serve as *t'ang-chang* or *chi-hu* (lit., "loom households"). With public funds supplemented by their own contributions, the *chi-hu* purchased raw silk, distributed it to private weavers, collected the finished product, and delivered it to the factory.[49] Because this led to the exploitation of weavers by the rich families and local officials, it was officially abolished in 1651.[50] Under the new system of *mai-ssu chao-pan* (lit., "buy silk and recruit help"), the factories bought raw silk directly on the open market and then recruited artisans to come to the factory to weave it. There was a system of rental and licensing of looms called *ling-chi kei-t'ieh* (lit., "lend looms and give licenses"). Those artisans who accepted the looms were known as *chi-chiang* or *kuan-chiang* (loom artisans or government artisans). Often the government artisans in turn hired their own weavers, or farmed out the work to others, thus serving in effect as middlemen or contractors, as the *t'ang-chang* or *pao-lan jen* had done in the Ming. Although this system purported to be an innovation, in fact it came to resemble the very system it was supposed to replace.[51]

What replaced the Ming hereditary artisan system was not a factory wage system, but a system through which responsibility and work were increasingly contracted out to private weavers. As P'eng Tse-i has observed, it was simply another way by which the government maintained its control over weavers.[52] In fact, the distinction between public and private continued to be ambiguous. As E-tu Zen Sun has expressed it,

> It would appear . . . that the "government-registered" textile producers were actually serving a dual function. They were responsible for maintaining their own establishments as part of the private economy while at the same time serving as contractors for the Imperial Silkworks. . . . Thus, while the silk industry in Soochow, Hangchow and Nanking was a flourishing private enterprise, it was also, through most of the Ch'ing dynasty, directly linked to the government procurement program.[53]

The recruitment of weavers and laborers directly by the Imperial Silk Factories, or indirectly through licensed weavers, has been regarded as an example of the development of a free labor market. At Soochow there was a famous market for wage laborers. The first record of this is from the late seventeenth century, although the practice probably existed earlier.

> In the eastern section of the prefectural city everyone is practised in the weaving profession. . . . The artisans each have their specialties. Those artisans who have regular masters are paid daily wages. If for some reason they call for a substitute from those without masters, this is called *huan-tai,* or "shouting for a substitute." Those without masters gather by the bridge at early dawn to wait. The satin workers stand at the Hua bridge, the damask workers stand at the Kuan-hua monastery bridge, and the spinners of silk, who are called *ch'e-chiang,* stand at the Lin-hsi *fang.* . . . Several hundred form a crowd, and they all stretch their necks in anticipation. Like vagrants, they gather together, and after having rice gruel, they scatter and return [whence they came]. If the master-weavers should reduce their output, this group will have no source of food and clothing.[54]

A similar situation apparently existed in Nanking during the Ch'ing. And the Hai-yen gazetteer of the Kuang-hsu period

(1875–1908) shows the existence of a seasonal labor force, which would travel from Hai-yen as far as Hu-chou and I-hsing to be hired as weavers during the non-farming months. They were apparently rural workers seeking off-season employment.[55]

The extent to which such hiring practices could be called "free," however, is questionable since, in the first place, they were clearly related to the operation of the Imperial Silk Factories insofar as the master-weavers served a dual public-private function. Moreover, the hiring of workers and artisans was subject to guild regulations. The daily labor markets were held near or on the premises of the guild and were regulated by it.[56] At Soochow, where the hiring of day workers was done at four different places, each location had a guild official who would supervise the proceedings—a system which often led to graft and corruption. Not until the late nineteenth and early twentieth centuries could it be argued persuasively that some guilds reflected working-class interests. Until then the guilds included both masters and workers.[57]

Examples of strikes and protests by weavers in the late Ming or early Ch'ing have been cited by historians as evidence of an emerging proletarian consciousness. In one well-known case in Soochow in 1601, weavers rose up against local officials after two years of rising taxes on looms and other economic burdens. A local weaver named Ko Ch'eng earned a name for himself in local history by accepting responsibility for a mob incident and going to jail for two years. Although the incident has been regarded as one of the first uprisings of handicraft workers, it is nevertheless clear that masters and workers reacted together and did not perceive their interests to be opposed. The strike was organized at the guild hall—another indication of the strength of the guild's influence on all weavers, both masters and workers.[58]

THE ORGANIZATION OF PRIVATE WEAVING

The growth of the commercial weaving sector provided the context for the evolution of new forms of production and large-scale commercial organization. Merchant enterprises known as *chang-fang, hao,* or *chuang,* operated putting-out systems through which they distributed raw materials and sometimes looms to weaving households, which wove the silk fabric and returned it in exchange for piece-rate wages, usually coupled with some kind of living allowance or stipend. This development has fascinated Chinese, Japanese, and Western historians. Liu Yung-ch'eng regarded the *chang-fang* (lit., "account house") as a capitalistic type of enterprise in which the relationship between *chang-fang* and weaver was that of employer and employed.[59] Yokoyama Suguru has called the *chang-fang* an example of the "highest stage of the development of commercial capital," having important similarities to the development of capitalism in the West.[60] Basing his work on Japanese scholarship, Mark Elvin concluded that "in the seventeenth century, small workshops or 'loom households' using hired labor were the norm, [but] . . . the putting out system in the urban silk industry probably first became of importance during the eighteenth century."[61]

In fact, the institution of putting out may have had its origins as early as the fifteenth century. According to P'eng Tse-i, it was not the practice of putting out that was new, but the terms *chang-fang,* or *sha-tuan chuang,* which were not used in Soochow until after the Taiping Rebellion. Whereas, in the early Ch'ing, the term *chi-hu* was applied to those wealthy households which contracted to get certain work done for the Imperial Factories, in the late Ch'ing, the term *chi-hu* applied to those who were employed to do the work, while the term *chang-fang* was applied to their employers. The whole notion of putting out, whether it was done for the factories or for commercial use, was also called *lan-chih,* or weaving on contract.[62] The work of the Japanese scholar Saeki Yūichi is consistent with this interpretation, saying that, while there is no direct evidence that organizations such as the *chang-fang* existed before the late Ming or early Ch'ing, all the indirect evidence suggests that they did.[63]

Even without direct evidence, it is possible to see the strong resemblance between late Ch'ing and twentieth-century private practices and those of the Imperial Factories of the late Ming period. By the late Ch'ing, Soochow, Nanking, and Hangchow had developed almost identical institutions, although there were some local differences in practice and terminology.[64] None of these institutions engaged directly in production, but they extended raw materials and looms to *chi-hu,* which in turn might hire artisans or workers.[65] At Soochow, there was a subcontracting system whereby the *chi-hu* would in turn assign some of its work to other weavers known as *tai-chi chi-hu* (lit., "substitute loom households"). The *chang-fang* were colloquially known as *ta-shu,* or "eldest uncle," the *chi-hu* as *erh-shu,* or "second uncle," and the *tai-chi chi-hu* were known as *san-shu,* or "third uncle."[66]

At Nanking, there was a *ch'eng-kuan* (lit., "responsible supervisor"), who was the middleman and guarantor between the account houses and the loom households. He was responsible for any losses of materials or equipment. Sometimes a further distinction was made between *nei* and *wai,* inner and outer, *ch'eng-kuan,* the latter having no direct contact with the account house. Without the *ch'eng-kuan,* no *chi-hu* could be contracted for directly. The *ch'eng-kuan* earned a commission for each bolt of silk produced.[67] Frequently, he worked simultaneously for several account houses or companies. He could earn a great deal of money this way, but the risks he assumed were also great.[68] The *ch'eng-kuan* apparently served some of the same functions as the "second uncle" in Soochow, but he was never directly engaged in production, while the "second uncle" probably was directly engaged in weaving.

As the term for "uncle" implies, the relationship between the account house and the loom household was a close and permanent one. The loom household was more or less permanently attached to one account house and would not freely switch its association to another. The relationship was sometimes spelled out in a written contract, and it conformed to the practices and regulations of the local guild.[69] The agreement entitled the loom household to place the account-house's chop on the finished product in order to identify it.[70] At Nanking, however, each loom household might weave on behalf of more than one account house. Each loom was designated for a particular *hao,* but actual ownership was vested in the household or workshop.[71] Both at Soochow and Nanking, payment was mostly by piece rate but, by the 1920s, some weavers in Nanking were paid wages.[72]

The *chang-fang* were characterized by their relatively large-scale operations. As Table 6 shows, in the early twentieth century, most controlled over one hundred looms but very few controlled more than three hundred. However, while each *chang-fang* may have controlled many *chi-hu,* each *chi-hu* remained a relatively small unit of production, on the scale of a workshop rather than a factory. In the late Ming and the very early Ch'ing, four or five looms had been typical of an urban weaving establishment, and twenty looms had been an absolute upper limit.[73] Although the official limitation on the number of looms per establishment had been removed because of a memorial submitted by Ts'ao Yin,[74] the scale of production seemed to have remained unchanged. Although the official regulations permitted a maximum of one hundred looms in each shop,[75] in 1902 the Maritime Customs reported that most Nanking weaving workshops still had only about four looms.[76] In Hangchow around 1904, according to a Japanese survey, there were about three thousand looms in the city, and each *chi-hu* had about two or three looms on the average. To have as many as ten looms was apparently exceptional.[77] In short, even in the twentieth century, the *chang-fang* were engaged in relatively large-scale financing, but not in large-scale production.

The *chang-fang,* however, indirectly controlled all the stages in the processing of the raw silk that preceded the actual weaving, as Figure 5 shows. It would buy warp from the *ching-hang,* or warp hong, and have it dyed and twisted. It would buy woof from the *ssu-hang,* or silk hong, and have it sorted and dyed.[78] Like the weaving itself, all these stages were closely controlled by the *chang-fang* and the guild organization behind

TABLE 6 Soochow and Nanking Account Houses
(*Chang-fang*)

	Number of Account Houses	Controlling Number of Looms
Nanking in the late Ching	3	500–600
	13	300–400
	9	100–200
	3	50–100
	6	20–40
	10	5–15
Total	44	7,829–8,129
Soochow in 1913	1	600
	6	300–400
	27	100–260
	15	50–90
	8	40–25
Total	57	7,681

Source: Yokoyama Suguru, "Shindai no toshi kinuorimonogyō no seisan keitai," *Shigaku kenkyū*, 105:55 (1968). For a complete table of all 57 Soochow *chang-fang*, together with information about their dates of establishment, ownership, number of workers, and output, see *Chung-kuo chin-tai shou-kung-yeh shih tzu-liao,* II, 428–430.

it.[79] Although we do not know exactly what procedures the *chang-fang* followed, some must have been able to exert considerable influence over the raw silk market simply by the scale of their operations. On the other hand, a *chang-fang's* functions, no matter how extensive, remained those of a middleman, handling transactions at various stages in the manufacturing process. Although the *chang-fang* imparted some degree of centralization to the otherwise highly decentralized structure of production, the units of production themselves, the warp hongs, the dyeing shops, and so forth, retained their small-scale and relatively separate identities.

The development of the *chang-fang* in the Ch'ing was closely

FIGURE 5 The Organization of the Silk Industry

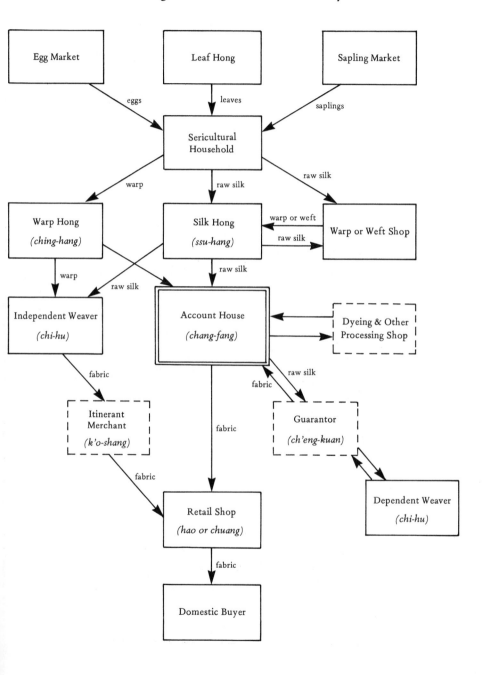

related to the expansion of interregional trade in silk goods. In fact the *chang-fang* were, first and foremost, merchant houses that got into the organization of manufacture in order to ensure an adequate supply of goods for their distant markets. There were also merchant houses, however, that engaged in inter-regional trade without getting into manufacturing, but simply purchased silk from independent weavers—sometimes through middlemen. In the first pattern, the large *chang-fang* of the major silk weaving centers organized weaving on a putting-out basis, and marketed the goods through branch stores in other cities. In the second pattern, *k'o-shang*, or itinerant merchants, were deputed by firms of their home localities to purchase and transport a given amount of silk fabric from independent weavers in the weaving centers.

At Nanking, in the twentieth century at least, the large *hao*, or firms, had branch shops in the major ports and cities, while the smaller *hao* might jointly open a store in another city or consign the sales of their goods to another *hao*. Alternatively, *k'o-shang* from various places resided at Nanking and purchased goods for their shops in the interior. These merchants often had to rely on middlemen to help them purchase fabrics.[80] Thus there continued to be a number of independent weavers who worked only for themselves.[81] At Hangchow, where only 30 percent of local goods were sold locally, similar practices prevailed. Large-scale *chuang*, having substantial amounts of capital and specializing in wholesale trade with other cities, would assign work to local loom households, but there were also smaller merchants who bought silk directly from independent weavers.[82] At Soochow, the smaller merchants who bought silk from independent weavers would also act as middlemen between itinerant merchants and local *chang-fang*.[83]

At other weaving centers, both patterns of production and marketing could be found. At Hu-chou, most weavers were independent in that they owned their own looms and bought their own raw silk. *Hu-ts'ou* was the principal fabric woven in this area, and the weavers sold it to the local *ts'ou-chuang*, or *ts'ou* dealer or shop, who in turn marketed it.[84] At Chinkiang, where weaving experienced a small boom at the end of the nineteenth century, the industry was organized as it was in the three large centers. There were several large *hao* which distributed raw silk to the weavers.[85] At Wu-ch'ing, both independent weavers and weavers attached to *ch'ou-chuang* existed.[86] At Sheng-tse, the itinerant merchants employed the services of middlemen called *ling-t'ou,* who went out into the countryside to purchase bolts of silk.[87] The *ling-t'ou* were specialized by types of fabric and, after years of experience, they built up a familiarity with the weavers in the countryside to whom they sometimes extended credit. In the 1920s and 1930s, there were one hundred thirty or forty such middlemen in Sheng-tse.[88]

THE SPROUTS OF CAPITALISM

Chinese and Japanese historians have expended a tremendous amount of scholarly energy in trying to find in the evolution of the Imperial Silk Factories into *chang-fang,* stages that correspond to the "putting-out" system and the development of "manufacture" in Europe.[89] In European economic history, the putting-out system has been regarded as a critical stage in the transition from feudalism to capitalism, because previously independent craftsmen fell under the domination of merchant-capitalists upon whom they became dependent for raw materials and, in many cases, for looms and other forms of fixed and working capital. From this stage, entrepreneurs went on to organize production in workshops, or what Marx called "manufacture," which was only one step removed from factory production under industrial capitalism. The transition from commercial capitalism to industrial capitalism witnessed the separation of the ownership of the means of production from the ownership of labor power.[90]

There is little reason to take issue with the view that the system of production organized by the *chang-fang* in the late Ch'ing bore striking resemblance to putting-out systems that emerged as early as the thirteenth century in continental

Europe,[91] and in the late sixteenth and early seventeenth centuries in the woolen industry in England.[92] Although differing according to time and location, these putting-out systems shared common characteristics. Most of them originated when merchants engaged in interregional or overseas trade sought to ensure themselves of a steady supply of manufactured goods, in many cases textiles, by organizing production, first in the cities and later in the countryside, where they could circumvent urban guild restrictions and find cheaper labor. If the raw materials came from a distant source, such as in the Italian silk industry,[93] or in the Wiltshire woolen industry,[94] the merchant was able to use his financial resources to ensure a supply of raw materials. The merchant-capitalist formed the critical link between raw materials, workers, and the wider export market.[95]

The preoccupation with comparing these Western developments with the *chang-fang* and related institutions, however enlightening it may be in some respects, is based on two fundamentally mistaken notions. The first is that putting out represented a new stage of development in Chinese economic history. In fact, however, the case of the Chinese silk weaving industry shows quite clearly the evolution of the practices of the *chang-fang* from the practices of the Imperial Silk Weaving Factories. Just as the *t'ang-chang* or *pao-lan jen* had agreed to fulfill certain weaving obligations for the Imperial Silk Factories, so too did the "second uncles," *ch'eng-kuan*, or other middlemen undertake the responsibility for procuring a certain amount of silk for the *chang-fang*. The basic principle of putting out was that of contracting work. The person who contracted to do the particular task assumed the responsibility for having the desired goods produced by others by the specified time. He guaranteed the results. The notion that the putting-out system was a new institution that did not fully develop until the mid-Ch'ing implies that the private weaving sector developed separately from the government weaving industry. In fact, however, public and private sectors were intertwined, and their institutional features were remarkably similar.

The second mistaken notion is that putting out would necessarily lead to a higher stage of economic organization, namely,
capitalism. Too often it is assumed that development of the
English cotton industry from a putting out system to factory
production constituted a universal and necessary pattern. In
fact it was the English historical experience that was exceptional and requires explanation. Putting out had existed in one
form or another for many centuries in Europe without generat-
ing an industrial revolution or a full-fledged capitalist system of
production. In Europe, as in China, a mature commercial network may have inhibited the development of capitalism. As
long as the merchant-capitalist could earn a good profit by
financing trade, there was no incentive to get involved in
production. Maurice Dobb has suggested that this was why
the putting-out systems in Italy, Germany, and the Netherlands
did not develop into a new stage of capitalistic production:

> It would seem as though the very success and maturity of merchant
> and money-lending capital in these rich continental centres of
> entrepot trade, instead of aiding, retarded the progress of invest-
> ment in production; so that compared with the glories of spoiling
> the Levant, or the Indies or lending to princes, industrial capital was
> doomed to occupy the place of a dowerless and unlovely younger
> sister.[96]

Dobb's work on capitalism has sparked a lively debate among
European Marxist historians about whether Marx himself did
not regard putting out as a form of merchant capitalism that
inhibited the transition to industrial capitalism.[97]

Only in the English cotton industry did very special circumstances combine to bring forth technological innovations that
triggered a revolution in the mode of production in the
eighteenth century. First, a vastly expanded domestic and
foreign market exacerbated the imbalances between the spinning
and weaving sectors—where it normally took five or six spinning
wheels to provide enough cotton for the operation of one
loom. The invention of the spinning jenny and later Arkwright's

water frame and Crompton's mule permitted a faster rate of spinning. At the same time their size necessitated factory production. Given the size of the labor force in England, it would have been impossible to expand production within the confines of the putting-out system without this organizational and technological breakthrough.[98]

In the Chinese cotton industry, the putting out system itself was not fully developed before the introduction of machine-spun yarn in the late nineteenth century. Although there was apparently some modified putting out practiced in some areas, in which the finished product would be exchanged for raw materials, on the whole, domestic production remained atomized.[99] Mark Elvin has attributed this to "the excellence of the market mechanism [which] made it unnecessary for cotton cloth merchants to become directly involved in production. They could hold almost all their capital in relatively liquid form as working capital and avoid tying it up as fixed capital." He explains that the network of urban settlement was so dense in Kiangnan, a center for cotton production as well as silk, that, "if a peasant with virtually no working capital could have daily access to a market for his or her materials, there was no place for a putting-out system."[100] Moreover, in China there was neither the rapidly expanding foreign market for cotton, nor the pressure of labor scarcity to create forces for change.

In the silk weaving industry, however, both the raw materials and the large looms required a greater outlay of capital than either cotton weaving or silk reeling. The development of *chang-fang* was probably necessitated by the expansion of interregional and foreign trade during the eighteenth century.[101] There was not, however, any further economic pressure to develop a factory system, since the great surge in foreign demand after 1842 was largely for raw silk and not for silk fabrics. Moreover, the social and political context of putting out in China was different from that of Europe in general, and eighteenth-century England in particular. European history saw the struggle of so-called independent craftsmen, organized in urban guilds, against local authorities and merchants alike. The putting-out system was an effort to by-pass the authority of the guilds by organizing rural labor. But in China, cities were not corporate entities separated from the countryside by law or custom, but were part of the central administrative system. Guilds worked hand-in-hand with local authorities to conduct and regulate commerce. Moreover, they did not represent the interests of craftsmen as distinct from merchants, but rather represented the interests of both together. Since the development of the *chang-fang* was closely associated with the spread of the guild system in the late Ch'ing, they can hardly be seen as an expression of "free" enterprise.

To regard the putting-out system as a "sprout of capitalism" is to apply alien categories of explanation to Chinese institutions that actually served a different purpose. Wherever putting out existed in China—whether it was in its old form under the Imperial Factories or its new *chang-fang* form—its fundamental purpose was to contract out work in a production process that was too complex to handle alone and to make a middleman, or series of middlemen, responsible for seeing that it got done. In the cotton-calendaring industry, a well-studied example of putting out in China, a *pao-t'ou* would serve as the middleman between the cotton merchant and the calendaring workshops; he was the guarantor.[102] *Pao*—"to contract," or "to guarantee certain achievement"—was a fundamental concept which pervaded all such economic relationships in China.[103] In a complex process, such as the silk weaving industry, the networks of *pao* could be quite lengthy. Financial management and production remained separate. The ultimate purpose was the diffusion of risk and responsibility among several different people, rather than their concentration in the hands of one entrepreneur. In short, it was totally contrary to the so-called spirit of capitalism.

THREE

The Silk Export Trade

The opening of the treaty ports after 1842 marked a new phase in the ancient and colorful history of the Chinese silk export trade. Whereas the traditional silk trade, from the Han through the Ch'ing, had been conducted over difficult overland and maritime routes, often through merchants of a third nationality, the modern export trade with the West served more demanding markets, which were more directly accessible through modern transportation and communication. Trade took on an even more competitive character as Japan entered the international silk market in the late nineteenth century. Gone were the days when Chinese silks were rare, exotic, and worth their weight in gold.[1] Gone were the days when Japanese daimyo, Portuguese Jesuits, and Mexican adventurers vied for shares of the lucrative silk trade. Gone also were the days when the Imperial government presumed to control the direction and the size of trade in this precious commodity.

FOREIGN TRADE BEFORE THE
TREATY PORTS

The export of Chinese silks during the Ming and Ch'ing, before 1842, fell into four major categories. First there was the trade with Central Asia and Russia. During the Ming and the Ch'ing silk continued to be, as it had in ancient times, an important item in the tribute trade. Silk cloth and garments held a prominent place in the list of gifts regularly presented by the court to foreign envoys. They also were a form of payment to the Mongols for the purchase of horses. Although it is impossible to determine from the records of these missions the total amount of silks exchanged in this manner, their relative importance must have declined over this period. Apparently some Mongol princes requested commutation of their silk gifts into cash payments, because they found the quality of the silk to have deteriorated.[2] Similarly in the Russian caravan trade of the eighteenth century, significant quantities of silk were traded, but the Russians complained about lowered quality, and cottons assumed a larger share of the total trade.[3]

The second category of foreign trade was that with Southeast Asia. Since this trade involved many ports of trade and many different nationalities—Chinese, Japanese, Portuguese, Spanish, Dutch, and others—it is even more difficult to get a sense of its aggregate size. More details, however, are known about the third category—the silk trade conducted by the Portuguese between Macao and Japan. During the fifteenth and early sixteenth centuries, there had already been a substantial direct trade in silk between China and Japan—both as part of the illegal pirate trade along the China coast and as part of the tally trade licensed by the Chinese and Japanese governments in an attempt to curb piracy. After they acquired Macao in 1557, the Portuguese dominated the silk trade between China and Japan for almost a century. With the reunification of Japan in the late sixteenth century, both Toyotomi Hideyoshi and Tokugawa

Ieyasu took pains to encourage the imports of silk; in fact, their desire to have this trade was one of their chief motives for tolerating the presence of Portuguese missionaries.[4] The Jesuits themselves derived financial as well as spiritual benefits from this trade, as they held a share of the cargo of the "Great Ship" which found its way to Japan almost annually.[5]

This trade was extremely profitable as there was a strong demand for silk in Japan, and Japanese sericulture was not yet well-developed. It was said that silk purchased for 80 to 90 ducats per picul in Canton could sell for 140 to 150 ducats in Japan.[6] There are some scattered records that show that exports reached 2,000 piculs a year, or 3,000 piculs in an exceptionally good year.[7] Since these records refer only to the tally trade in raw silk, we can conservatively estimate that the annual volume of silk was double that quantity, or 4,000–6,000 piculs a year. This trade resulted in such an outflow of silver specie from Japan that the Tokugawa Bakufu tried to limit it to 750 piculs a year, but without much apparent success. Although the silk trade continued through Dutch and Chinese merchants after the expulsion of the Portuguese from Japan in the mid-seventeenth century, by the end of the century, the Bakufu began to adopt a policy of actively encouraging the development of sericulture within Japan to reduce Japanese dependence on Chinese silk.[8]

Of even greater economic importance was the fourth category of silk export trade, the triangular trade between China, the Philippines, and Spanish America from the mid-sixteenth through the eighteenth century. Chinese raw silk, mostly from Chekiang, but sometimes from Kwangtung and even Szechwan, as well as all varieties of silk fabrics and apparel, were shipped by Chinese merchants from Canton or Chang-chou to Manila, where the Spanish had installed themselves in 1571.[9] From Manila, an average of one to four Spanish galleons would sail to Acapulco each year, carrying Chinese silks as their major cargo to an eager Spanish American market. A weaving industry employing 14,000 persons grew up in Mexico simply to handle the raw silk sector of this trade. It was said that Chinese silks could be found on the backs of even ordinary persons and on the altars of churches all over Spanish America.[10] The profits earned by Chinese and Spanish traders were reported to be phenomenal; some Spanish traders were said to have realized tenfold gains on their investments. According to one source, two to three million pesos' worth of Chinese silks were imported to Mexico each year. Since the price of silk may have reached 500 taels per picul in Mexico, at the rate of 0.8 tael per peso, we may calculate that at least 3,000–5,000 piculs of raw silk entered Mexico yearly, and probably a great deal more. According to other estimates, however, 10,000–12,000 bales of Chinese silk were shipped during one year, which was probably the equivalent of 8,000–10,000 piculs of raw silk.[11] In exchange for these silks, a vast quantity of Mexican and Peruvian silver bullion was shipped back to Manila and eventually found itself in Chinese hands. During the seventeenth and eighteenth centuries, when China's own silver mines were depleted, this influx of silver had a profound effect on its domestic economy—on the one hand accelerating the trend toward monetization, and on the other, contributing to a steady rise in price levels during the eighteenth century.[12]

With the decline of Portuguese and Spanish maritime supremacy, the Dutch, and later the English, began to play a more important role in Asian trade. Although Amsterdam was briefly a major European entrepot for silk, on the whole the Dutch were more involved in trading spices and other commodities.[13] For the British, however, silk had greater importance than tea until the mid-eighteenth century.[14] Although the silk trade continued to grow in absolute terms, with the Commutation Act of 1784 by which the British Parliament greatly cut the duties on tea imports, silk was overshadowed by tea, which became the major commodity traded by the East India Company in China, eventually representing almost the totality of its business there. Between 1775 and 1785, silk represented about 31 percent of the exports of the East India Company from China but, between 1785 and 1795, its share had dropped to less than 10 percent.[15] After 1825, the East India Company stopped shipping silk altogether, leaving the trade to private merchants.[16] After the abolition of the company's monopoly in 1833, there was a dramatic increase in the quantity of both tea and silk shipped from Canton. While the average amount of raw silk exported annually from Canton between 1830 and 1833 had been about 4,300 piculs, between 1834 and 1837 the annual average rose to 10,000 piculs. With the outbreak of the Opium War, however, exports again fell to an annual average of 2,500 piculs in the 1838–1842 period.[17]

The British demand for Chinese raw silk, as modest as it was in the eighteenth century, stemmed in large part from the establishment of Thomas Lombe's famous silk-throwing mill at Derby—built allegedly with Italian designs smuggled to England by his brother. This mill employed hundreds of workers and set the pattern for the cotton spinning factories of future years.[18] From this time, British imports emphasized raw silk rather than thrown and woven silks.[19] China, however, was not the sole source of raw silk in the late eighteenth century. England also imported some silk from Italy, Turkey, and Bengal, the quantity from the latter source exceeding that imported from China.[20]

For the British, the silk trade at Canton was subject to all of the frustrations characteristic of the "Canton system." Orders for tea and silk could be placed only through the hong merchants, who were all-powerful when it came to the question of quotas and surcharges. The advance deposits required for silk were especially large; 90 percent of the cost was required as opposed to only 50 percent for tea. Orders had to be placed months in advance. In the early days of the trade, when there were only four ships per season, and only 300–400 piculs involved, delivery could be made in a hundred days, but later when the quantities increased to thousands of piculs, orders had to be placed in March for the following winter.[21]

During the eighteenth century the price of raw silk rose very steeply, particularly in the second half. Sometimes it was impossible for the supercargoes of the East India Company to fill their orders when the price was too high.[22] Chinese authorities were very alarmed by the price trend, fearing its impact on the ability of the Imperial Silk Weaving Factories to procure their usual supply of raw silk. Attributing the rise in prices to an increase in foreign trade, the court sought to impose tighter restrictions on exports. In 1759 an Imperial edict enjoined officials of Kiangsu and Chekiang to enforce strictly the regulations for the export of raw silk, which in recent years, it said, had been disregarded by "treacherous merchants who sought illicit profit [and] sold secretly to foreign countries."[23] In 1762, however, a local official observed that the enforcement of the restrictions had not brought down the price of silk, remarking that "this phenomenon is probably due to the daily increase in our population. The price of goods in limited supply will inevitably be high." An Imperial edict in response to his memorial established new quotas for the export of raw silk. Each ship was permitted to carry 5,000 catties of raw "local" silk, which was presumably Kwangtung silk, and 3,000 catties of raw silk from second-generation silkworms, but the export of Hu-chou silk from first-generation worms, as well as silk and satin fabrics, was still prohibited.[24] Two years later the quotas were further relaxed, apparently at the urging of some provincial officials who had been implored by Dutch and British merchants to seek a change in the regulations. These officials argued that no real harm was done by exports, since the best silk was reserved for Imperial use in any case. At Hangchow, Chia-hsing, and Hu-chou, the prefects reported that local opinion held that the relaxation of export restrictions would be advantageous.[25] New quotas were imposed, but these were changed from year to year. By 1784, the quota allowed Western ships had been raised to one hundred piculs per ship.[26] It was clear, however, that all these restrictions were more often honored in their breach, frequently being circumvented by Western vessels.[27]

Despite these measures, the price of the best raw silk at Canton more than doubled during the course of the eighteenth century, as Table 7 shows. The English traders referred to this as "Nanking" silk, although it was actually raw silk from the Hu-chou area. "Local" silk produced in the Canton delta area was also exported, but its price was substantially lower and more stable than that of Hu-chou silk.[28] It seems, however, that Ch'ing officials were wrong in assigning the cause of the price rise to foreign trade. Although the British trade in Chinese raw silk had grown in the latter half of the eighteenth century, its size was small in comparison to the Portuguese or Spanish trade which had preceded it in the seventeenth and early eighteenth century. Although there is no way to make a reasonable estimate of the size of the domestic market for raw silk, it seems unlikely that even the Portuguese Macao-Nagasaki trade of perhaps 6,000 piculs a year, or the Spanish-American trade of perhaps 10,000 piculs, not to mention the eighteenth-century English trade of a few hundred piculs, came close to exerting pressure on the available supply. If it had, the price of silk should have started to rise in the seventeenth century.

A more plausible explanation was contained in the observations of the memorialist cited above. The growth in China's population, together with increasing peace and prosperity, had stimulated the domestic markets for silk and forced the price upward. In fact, the rise in silk prices was entirely consistent with the overall price inflation of the eighteenth century. It is estimated, for example, that the price of rice, a leading indicator, rose about 125 percent for the century as a whole.[29] On the other hand, since one of the major causes of the general inflation, in addition to demographic factors, was the massive influx of silver from Spanish America, in a very general and indirect sense it could be said that the silk export trade did affect the price level.[30]

The greater sensitivity of silk prices to internal rather than external pressures gives us a good indication of the extensive-

TABLE 7 Raw Silk Prices at Canton, 1702–1799

Year	Price per picul (taels)	Index	Average price (taels)
1702	132	100	
1703	140	106+	
1704	100	76+	
1722	150	113+	
1723	142–144	100.8	143
1724	155	117+	
1731	155–159	118+	157
1750	175	131.8+	
1755	190–195	145.4+	192.5
1757	225–250	180+	238
1763	245	185+	
1765	269	203.7+	
1770	300	227+	
1775	275–277.5	209+	276
1783	275	207+	
1784	310	234+	
1792	312	236+	
1793	255	193+	
1799	270	204+	

Source: E-tu Zen Sun, "Sericulture and Silk Textile Production in Ch'ing China," in W. E. Willmott, ed., *Economic Organization in Chinese Society* (Stanford, 1972), p. 91, as adapted from Ch'üan Han-sheng, "Mei-chou pai-yin yü shih-pa shih-chi, Chung-kuo wu-chia ko-ming ti kuan-hsi," *Chung-yang yen-chiu yuan li-shih yü-yen yen-chiu so ch'i-k'an,* 28:534 (1957).

ness of production. European observers in China were universal in their opinion that the supply of Chinese silk was enormous. In 1736 the Jesuit priest, J. B. Du Halde, wrote in his impressions of China:

> *China* may be called the country of silk, for it seems to be inexhaustible, supplying several nations in Asia and Europe, and the Emperor, the Princes, the Domesticks, the Mandarins, the men of Letters, women, and all in general whose circumstances are tolerable wear Garments of Silk, and are clothed with Satin or Damask; there are none but the meaner People and Peasants that wear blue Calicoes.[31]

Even after the opening of Shanghai as a treaty port, when raw silk exports quickly expanded from about 5,000 piculs in 1845 to 68,000 piculs in 1858, as shown in Table 8, prices dropped rather than increased.[32] Again, foreigners observed that the Chinese supply was large enough to cushion the impact of the sharp rise in foreign demand. S. Wells Williams observed:

> The quantity produced to supply the native consumption is so enormous, that not withstanding the vast increase in export during the past ten years, the average of prices is lower than when the export was but one-fourth of its present amount. The silk grower looks to the home market for fixing the value of his produce, and prices range according as that demand is active or dull; little or no effect being produced by the foreign exportation, except among the speculative holders at the ports.[33]

The English botanist, Robert Fortune, whose reports on silk and tea districts were authoritative, observed that the fact that exports of silk could expand so rapidly after 1842 was clear evidence of

> the enormous quantity, which must have been on the Chinese market before the extra demand could have been so easily supplied. But as it is with tea, so it is with silk,—the quantity exported bears but a small proportion to that consumed by the Chinese themselves. The 17,000 extra bales sent yearly out of the country have not in the least affected the price of raw silk or silk manufactures. This fact speaks for itself.[34]

Indeed prices were so clearly determined by the domestic market that Western buyers often found it hard to fill their orders. The Shanghai representatives of the American company of Augustine Heard reported in the late 1850s that "dealers are very firm in consequence of the high prices in the interior," or "the views of the buyers are so much lower than the holders, that we can report only a settlement of 1,000 bales."[35]

The treaty system, which was introduced after 1842, brought an end to the Canton system, which the British had found so restrictive, and opened five coastal ports to foreign trade. The

TABLE 8   Raw Silk Exports from Canton and Shanghai,
1843–1858 (in piculs)

Year	Canton	Shanghai	Total	Shanghai as % of total
1843	1,430	—	—	—
1844	2,083	—	—	—
1845	5,430	5,146	10,576	48.7
1846	2,843	12,157	15,000	81.0
1847	960	16,941	17,901	94.6
1848	—	14,507	—	—
1849	849	12,190	13,039	93.5
1850	3,444	13,794	17,238	80.0
1851	1,927	16,505	18,432	89.5
1852	2,839	22,461	25,300	88.8
1853	3,662	46,655	50,317	92.7
1854	—	43,386	—	—
1855	—	44,969	—	—
1856	—	63,357	—	—
1857	—	47,989	—	—
1858	—	68,776	—	—

Source: Adapted from Hosea Ballou Morse, *The International Relations of the Chinese Empire* (London, 1910–1918), I, 366. 1 bale = 0.8 picul.

major structural change for the silk trade was the shift of activity from Canton to Shanghai. The proximity of Shanghai to the silk-growing districts of Chekiang was in large measure responsible for its spectacular growth as a treaty port, in contrast to Ningpo, Foochow, and Amoy.[36] The easy system of water transport from these districts to Shanghai greatly reduced the costs of internal transport. Furthermore the duties on silk were kept very low in the treaty revision of 1858, at only ten taels per picul, instead of the general tariff rate of 5 percent ad valorem.[37] Instead of dealing with the hong merchants, Western trading companies at Shanghai dealt with compradores, usually Chekiang natives, with whom they placed their orders.

These middlemen went into the silk districts to negotiate purchases, while the company representatives remained at Shanghai.[38] Although the Western merchants were still dependent on intermediaries to help them reach producers, no longer were the latter licensed and controlled by the central government.

GENERAL TRENDS, 1870–1937

For the purpose of understanding the basic trends in the silk export trade, 1870 is a more useful transitional date than 1842. First, the Taiping Rebellion (1850–1864) did great damage to the sericultural districts as well as the urban silk weaving industry when its forces struck the lower Yangtze valley region in the early 1860s. The extent and nature of this damage was believed by Westerners to have been very severe and to have permanently affected the silk trade in that area. In the next chapter we shall discuss whether this was in fact the case, but at the very least the rebellion did temporary harm to both rural and urban sectors of the silk industry. By 1870 recovery was just under way. Second, it was only after 1870 that the Imperial Maritime Customs Service, an agency through which foreigners collected customs duties on behalf of the Imperial government, began to adopt uniform procedures for the collection of trade statistics. After 1873 trade figures were given in a common denomination, the Haikwan or Customs tael, for the first time.[39]

In the period from 1870 to 1930, China's exports of both raw silk and silk fabrics increased steadily. As shown in Table 9, the total value of silk exports increased more than sevenfold between 1870 and 1928, from 23,518 HK taels to 169,347 HK taels. Exports of raw silk more than tripled in volume. In 1870, 49,000 piculs of raw silk were exported; by the 1890s annual raw silk exports averaged over 100,000 piculs; and by 1928, about 180,000 piculs were exported. In the same period, the volume of silk fabric exports increased about eightfold, from about 4,000 piculs in 1870 to about 33,000 piculs in 1928. Although the portion of the total silk export trade represented by fabrics was only about 8 percent by value at the beginning of this period, from the 1880s fabrics held between 15 to 25 percent.[40] Silk fabrics, however, never played as important a role in the modern export trade as raw silk.

Silk and tea represented the major portion of China's export trade in the nineteenth century, but the tea export trade declined rapidly in the face of international competition. By 1887, silk had overtaken tea as China's major export, and by 1898, the value of the silk export trade was about double that of the tea trade, as shown in Table 10.[41] Although the value of the silk trade continued to rise, and although silk and silk products continued until the late 1920s to be China's most important export commodity, nevertheless, the relative share of total trade represented by silk declined gradually throughout the period under consideration, from over 40 percent in the late nineteenth century, to about 16 percent in 1930. This represented a general trend in the twentieth century toward a diversification of Chinese exports and away from reliance on one or two key commodities.

About two-thirds of the raw silk export trade went through the port of Shanghai after 1870, as shown in Table 11. Most of the other third went through Canton. Shanghai's exports, however, included reexports from north China, as well as silk produced in Chekiang and Kiangsu. By 1925, about 11 percent of Shanghai's raw silk exports consisted of yellow silk from Szechwan, and 15 percent Tussah silk from Manchuria.[42] Canton's raw silk exports depended mostly on the growth of its steam filature industry in the late nineteenth century, as Table 12 shows. It also had a strong piece-goods industry, which had a steady foreign market, particularly in the nineteenth century.[43]

There were many types and grades of raw silk exported from China. Of these white silk was of the highest quality and had

TABLE 9 Raw Silk and Silk Fabric Exports from China, 1859–1937

	Raw Silk		Silk Fabric		Total by	Silk Fabrics
Year	*Weight* (1,000s of piculs)	*Value* (1,000s of taels)	*Weight* (1,000s of piculs)	*Value* (1,000s of taels)	*Value* (1,000s of taels)	*as %* *of Total* *Value*
1859	52	17,549				
1860	63	21,810				
1861	51	17,768				
1862	61	23,857				
1863	30	10,121				
1864	24	9,455				
1865	41	16,405				
1866	31	14,226				
1867	45	16,371	4	2,172	18,543	11.7
1868	57	25,109	4	1,947	27,056	7.2
1869	48	19,583	3	1,695	21,278	8.0
1870	49	21,641	4	1,877	23,518	8.0
1871	60	25,469	4	2,353	27,822	8.5
1872	65	27,901	5	2,607	30,508	8.5
1873	61	28,289	5	2,203	30,492	7.2
1874	75	22,123	6	2,375	24,498	9.7
1875	80	20,107	6	4,023	24,130	16.7
1876	79	30,908	6	3,986	35,894	11.1
1877	59	17,623	6	4,432	22,055	20.1
1878	67	19,830	7	4,507	24,337	18.5
1879	81	23,006	7	4,499	27,505	16.4
1880	82	22,990	8	5,422	28,412	19.1
1881	66	20,124	7	4,612	24,736	18.6
1882	65	17,335	7	3,396	20,731	16.4
1883	65	17,470	8	4,023	21,493	18.7
1884	68	16,457	9	4,427	20,884	21.2
1885	58	13,570	10	4,556	18,126	25.1
1886	77	19,210	12	6,754	25,964	26.0
1887	79	20,741	14	6,723	27,464	24.5

TABLE 9 (continued)

Year	Raw Silk		Silk Fabric		Total by Value (1,000s of taels)	Silk Fabrics as % of Total Value
	Weight (1,000s of piculs)	Value (1,000s of taels)	Weight (1,000s of piculs)	Value (1,000s of taels)		
1888	77	20,070	16	7,894	27,964	28.2
1889	93	24,801	15	7,175	31,976	22.4
1890	80	20,626	11	5,320	25,946	20.5
1891	102	26,030	13	6,465	32,495	19.9
1892	101	27,323	16	7,372	34,695	21.2
1893	94	25,788	18	8,253	34,041	24.2
1894	99	29,219	19	8,415	37,634	22.4
1895	111	34,576	24	11,331	45,907	24.7
1896	88	28,710	21	9,723	38,433	25.3
1897	117	40,993	20	10,095	51,088	19.8
1898	109	40,781	20	10,044	50,825	19.8
1899	148	65,245	18	9,893	75,138	13.2
1900	97	36,555	18	9,028	45,583	19.8
1901	129	46,368	21	10,227	56,595	18.1
1902	120	62,128	20	9,652	71,780	13.4
1903	95	51,211	20	13,785	64,996	21.2
1904	125	61,327	21	11,764	73,091	16.1
1905	106	53,425	15	9,939	63,364	15.7
1906	111	56,048	16	9,754	65,802	14.8
1907	116	67,891	21	12,927	80,818	16.0
1908	129	62,128	23	13,728	75,856	18.1
1909	130	64,029	29	17,892	81,921	21.8
1910	139	71,546	30	17,998	89,544	20.1
1911	130	64,935	28	17,051	81,986	20.1
1912	158	67,691	38	16,104	83,795	19.2
1913	149	73,510	34	20,100	93,610	21.5
1914	109	55,561	28	15,991	71,552	22.3
1915	143	69,079	41	21,558	90,637	23.8
1916	122	78,262	39	20,020	98,282	20.4
1917	126	79,149	30	17,230	96,379	17.9

TABLE 9 (continued)

Year	Raw Silk		Silk Fabric		Total by Value (1,000s of taels)	Silk Fabrics as % of Total Value
	Weight (1,000s of piculs)	Value (1,000s of taels)	Weight (1,000s of piculs)	Value (1,000s of taels)		
1918	125	74,682	35	18,911	93,593	20.2
1919	166	102,549	40	23,261	125,810	18.5
1920	104	68,154	38	24,318	92,472	26.3
1921	151	112,143	43	30,274	142,417	21.2
1922	143	137,217	31	23,631	160,848	14.7
1923	138	138,916	29	24,548	163,464	15.0
1924	131	108,059	27	22,301	130,360	17.1
1925	168	140,358	31	23,203	163,561	14.2
1926	169	144,736	39	30,858	175,594	17.6
1927	160	128,706	33	25,171	153,877	16.3
1928	180	145,443	33	23,904	169,347	14.1
1929	190	147,681	29	21,033	168,714	12.5
1930	152	109,181	30	19,565	128,746	15.2
1931	136	84,681	34	24,413	109,094	22.4
1932	78	32,932	22	15,710	48,642	32.3
1933	77	48,246	20	20,789	69,035	30.1
1934	54	23,519	19	16,459	39,978	41.2
1935	76	35,679	17	12,006	47,685	25.2
1936	63	36,713	17	11,593	48,306	24.0
1937	69	45,866	17	11,150	57,016	19.6

Sources: For 1859–1867: T. R. Banister, "A History of the External Trade of China, 1834–1881," in China, Maritime Customs, *Decennial Reports, 1922–1931* (Shanghai, 1933), I, 91. Banister notes, "The quantity for 1859 contains an estimate only of the export from Canton. The values from 1859–1866 inclusive have been computed but being calculations only and not records should be taken with some reservation."

For 1867–1937: Adapted from Liang-lin Hsiao, *China's Foreign Trade Statistics, 1864–1949* (Cambridge, Mass., 1974), pp. 102–116. Raw silk includes the following categories from the returns of the Maritime Customs: 1) White, not rereeled, not steam filature, 2) White, rereeled, 3) White, steam filature, 4) Yellow, not rereeled, not steam filature, 5) Yellow, rereeled, 6) Yellow, steam filature, 7) Wild, filature, and 8) Wild, not filature. It excludes Hsiao's Column 39, "Raw Silk, Miscellaneous," because inclusion of cocoons of various types and waste silk, which are subsumed under this category, greatly inflates the weight figures for each year. These same total raw silk figures are found also in Tuan-liu Yang and Hou-pei Hou, (C. Yang and H. B. Hau), comp., *Statistics of China's Foreign Trade during the Last Sixty-five Years, 1864–1928* (Nanking, 1931), p. 41.

TABLE 9 (continued)

Silk Fabric includes 1) Silk Piece Goods, and 2) Silk, Shantung Pongees. It excludes Hsiao's Column 43, "Silk, Embroideries," and Column 44, "Silk Products, Miscellaneous."

Note: Before 1868, values are given in Shanghai taels. Starting in 1868, figures are given in Haikwan taels.

---

the largest foreign market. The most famous type of white silk was called Tsatlee, named for the Ch'i-li district of the Nan-hsun *chen* of Hu-chou prefecture, in the heart of the sericultural region of north Chekiang. After 1842, Tsatlees experienced a huge surge in popularity. However, because they were reeled domestically, Tsatlees were often of uneven quality. In order to make them more competitive with machine-reeled silk, Tsatlees were rereeled for the foreign market in the twentieth century, appearing for the first time in the Customs statistics in 1910. Other types of domestically reeled silks from the Kiangnan area included "Kashings" from Chia-hsing prefecture, called "coarse and uneven" by one American expert, and "Haineen rereels" and "Wuchun rereels," from the districts

TABLE 10 Silk and Tea as Percentage of Total Chinese
Exports and Shanghai Exports, 1850–1930

| Year | Silk | | Tea | |
	China[a]	Shanghai[b]	China	Shanghai
1850	—	52	—	46
1860	—	66	—	28
1870	—	62	—	32
1880	38	38	46	48
1890	35	34	31	30
1900	31	30	16	16
1910	26	26	9	10
1920	19	20	2	4
1930	16	14	3	4

Sources: [a]*Decennial Reports, 1922–1931,* I, pp. 120, 190.
[b]Rhoads Murphey, *Shanghai: Key to Modern China* (Cambridge, Mass., 1953), p. 119, based on customs returns and other estimates.

TABLE 11   Raw Silk Exports from Shanghai, 1870–1925

Year	Shanghai Exports[a] (in piculs)	Total Chinese Exports[b] (in piculs)	Percentage of Total Represented by Shanghai
1870	30,482	49,160	62.0
1875	55,965	79,914	70.0
1880	69,685	82,201	84.8
1885	44,690	57,984	77.1
1890	51,808	80,400	64.4
1895	76,639	110,621	69.3
1900	60,432	97,207	62.2
1905	63,299	105,919	59.8
1910	87,540	139,226	62.9
1915	95,822	143,097	67.0
1920	54,366	104,315	52.1
1925	120,503	168,017	71.7

Sources: [a]*Annual Trade Returns and Reports,* comp. China, Maritime Customs, annual, until 1920. After 1920, *Quarterly Returns of Trade.* All types of raw silk included, except waste silk and cocoons.
[b]Tuan-liu Yang and Hou-pei Hou, *Statistics of China's Foreign Trade,* p. 41.

of Hai-ning in Hangchow prefecture and Wu-chen in Hu-chou prefecture respectively.[44] There were also Taysaams, named for *ta-ts'an,* lit., "big silkworms," which were a type of coarse silk from Chia-hsing and Shao-hsing also popular abroad.[45] Beginning in the 1890s, these domestically reeled silks gradually were displaced in the export market by machine-reeled, or steam filature, silk. Appearing first in the Customs figures in 1894, steam filatures quickly reached 45 percent of all white raw silk exports by the end of the century, and over 80 percent by 1920, as shown in Table 13. Steam filature silk was sold at a much higher price than domestically reeled silk, as shown in Table 14, and thus greatly increased the value of silk exports.

Hand-reeled and machine-reeled white silks constituted the bulk of the raw silk export trade, but other types were also

TABLE 12   Steam Filatures as Percentage of Total White Silk
Exports: Shanghai and Canton, 1895–1930

Year	Shanghai	Canton
1895	11.7	85.2
1900	18.5	97.9
1905	36.5	94.6
1910	44.9	96.4
1915	98.5	81.1
1920	97.3	96.0
1925	99.4	98.5
1930	97.5	98.9

Source: Adapted from H. D. Fong, "China's Silk Reeling Industry: A Survey of Its Development and Distribution" *Monthly Bulletin on Economic China* 7.12:491 (December 1934), Table V. Fong's data present some difficulties because his figures for Shanghai do not include rereeled white silk, which greatly exceeded ordinary hand-reeled silk after 1910. If rereeled silk were added to his totals for hand-reeled silk, as it is in Table 13, the percentage figures for steam filature silks exported from Shanghai after 1910 would be considerably lower than suggested here. For figures on rereeled silk, see Liang-lin Hsiao, pp. 102–103.

important. Yellow silk, mainly from Szechwan and partly from Shantung, had a moderate-sized foreign market. After the turn of the century, wild silk, produced from silkworms that fed on oak leaves, became extremely popular in the United States and France. These silks from Manchuria and Shantung were called Tussah, or Tussores, and were woven into Pongees, which were also popular abroad.[46] The great merit of Pongees was that they could be washed easily and yet were no more expensive than good cottons.[47] Their sudden popularity in the 1910s through the 1930s was to a large extent responsible for the increase in silk piece-goods exports in this century. By 1921, the value of Shantung Pongee exports reached thirteen million HK taels, whereas all other types of piece-goods exports were valued at seventeen million HK taels.[48]

Waste silk, cocoons, and refuse cocoons were also exported. Although the quantities were not large, there was a steady

TABLE 13 White Raw Silk Exports, 1894–1932:
Domestic and Steam Filature Silk

	Domestically Reeled White Silk[a]		Steam Filature White Silk[b]		
Year	Quantity (1,000s of piculs)	Value (1,000s of HK taels)	Quantity (1,000s of piculs)	Value (1,000s of HK taels)	Percent of Total Quantity[c]
1894	69	22,973	4	2,324	5.5
1895	56	19,140	27	11,210	32.5
1896	38	13,286	27	11,126	41.5
1897	48	17,441	41	18,718	46.1
1898	44	17,687	41	18,103	48.2
1899	60	29,104	49	26,335	45.0
1900	32	14,523	35	16,039	52.2
1901	45	17,603	50	21,807	52.6
1902	37	20,620	51	33,372	58.0
1903	19	11,603	44	31,285	69.8
1904	34	19,582	47	28,526	58.0
1905	24	13,524	45	27,396	65.2
1906	27	16,485	46	29,614	63.0
1907	29	17,804	50	39,047	63.3
1908	32	17,714	49	32,318	60.5
1909	31	15,383	52	34,341	62.7
1910	31	15,786	64	42,701	67.4
1911	27	14,232	55	36,780	67.1
1912	43	18,377	57	34,377	57.0
1913	33	15,426	68	45,602	67.3
1914	16	8,363	54	37,384	77.1
1915	32	16,146	60	40,189	65.2
1916	20	11,823	66	53,771	76.7
1917	19	11,852	69	53,301	78.4
1918	18	11,125	61	47,770	77.2
1919	21	12,652	83	69,364	79.8
1920	11	7,641	54	46,376	83.1
1921	11	7,841	79	72,479	87.8
1922	16	12,008	84	99,058	84.0

TABLE 13 (continued)

| Year | Domestically Reeled White Silk[a] | | Steam Filature White Silk[b] | | |
	Quantity (1,000s of piculs)	Value (1,000s of HK taels)	Quantity (1,000s of piculs)	Value (1,000s of HK taels)	Percent of Total Quantity[c]
1923	16	12,118	70	92,972	81.4
1924	13	9,172	73	74,308	84.9
1925	18	12,142	90	96,775	83.3
1926	16	10,961	97	103,364	85.8
1927	18	11,785	92	90,018	83.6
1928	18	10,294	105	103,264	85.4
1929	25	13,419	110	106,516	81.5
1930	14	7,108	90	77,468	86.5
1931	8	3,732	68	51,823	89.5
1932	8	3,005	40	19,211	83.3

Source: This table is constructed from data in Liang-lin Hsiao, Table 3, pp. 102–103. Although steam filature silk had been exported from Canton much earlier, 1894 was the first year it appeared in the Maritime Customs records. Yellow and wild silk are excluded.

Notes: [a]This includes all white silk which was not reeled in steam filatures. After 1910, domestically reeled silk included rereeled silk. Therefore this column includes both columns 31 and 32 in Hsiao's statistical volume. The price of rereeled silk was of course significantly higher than that of silk not rereeled.
[b]This data is from column 33 of Hsiao's table.
[c]Steam filature exports as a percentage of total white silk exports.

European market for each. Western machinery was developed to process these materials rejected or discarded in the hand-reeling process.[49] In the 1920s, there were at least seventy different categories of waste silk alone.[50]

THE INTERNATIONAL SILK MARKET

In the transitional period between the Treaty of Nanking and the end of the Taiping Rebellion, most Chinese silk exports were shipped to Great Britain, where they were reshipped to France and other destinations in Europe.[51] It was not until

TABLE 14 Prices of White Raw Silk at Shanghai,
1862–1931
(Five-year averages in HK taels per picul)

Years	Domestic, not rereeled	Steam Filature
1862–1866[a]	407	
1867–1871	497	
1872–1876	404	
1877–1881	328	
1882–1886	294	
1887–1891	312	
1892–1896	319	
1897–1901	392	669
1902–1906	537	795
1907–1911	493	822
1912–1916[b]	440	744
1917–1921	355	865
1922–1926[c]	738	1,121
1927–1931	674	923

Sources: [a]1862–1911 figures adapted from *Silk: Replies from the Commissioner of Customs,* comp., China, Maritime Customs (Shanghai, 1917), p. 204.
[b]1912–1921 figures from *Decennial Reports, 1912–1921*, p. 436.
[c]1922–1931 figures from *Decennial Reports, 1922–1931*, p. 189. These refer to prices for all China, not just Shanghai.

after 1868 that the French began to purchase silk directly and the London middleman became less important.[52] The appearance of France as a major customer on the world silk market was a major factor in the subsequent development of the Chinese trade. France had long practiced a restrictive trade policy in order to protect its silk weaving industry at Lyon,[53] and its own sericultural output prospered in the early nineteenth century, reaching twenty-six million kilograms, or about 433,000 piculs of cocoons in 1853. But the next year a terrible plague affected the French silkworm crop, and output plummeted to four million kilograms or 67,000 piculs by 1865.[54] The disease spread to other parts of Europe and the Middle East. Although it was eventually brought under control by the techniques of Louis Pasteur, France became dependent on China for its supply of raw silk. Chinese silk was not only of high quality, but it could be imported at a cost often lower than that of raising silkworms in France. Sericulture never regained its former prominence in France, although the weaving industry remained strong. In the late nineteenth and early twentieth centuries, France was China's most important customer for raw silk, particularly Tsatlees. As Table 15 shows, by the 1910s, the United States began to rival and then overtake France as a customer for silks shipped from Shanghai. In the Canton export trade, by contrast, the American market was initially more important in the early twentieth century, but by the mid-1920s the French market reached almost equivalent proportions.[55]

The emergence of the United States as a major consumer of raw silk was the single most important development in the international silk market of the late nineteenth and twentieth centuries. By the twentieth century the United States was the leading customer for raw silk, and New York had become the leading international silk center.[56] The Silk Association of America reported that in 1916, for example, the United States had imported over thirty-one million pounds (about 233,000 piculs) of raw silk, or about 60 percent of the total traded internationally. By the late 1920s, the United States was importing 450,000 piculs of raw silk annually.[57]

On the supply side, the most dramatic change in the international market was the entrance of Japan as a major producer and exporter of raw silk. In the 1870s and 1880s, when the Japanese silk industry was still in a formative stage, Americans still preferred Chinese silk, but purchased Japanese silk when the Chinese price went too high. In those days American buyers complained that Japanese silk was irregular and inferior,[58] while also complaining of dishonest Chinese business practices.[59] The Japanese attention to the needs of foreign buyers, however, and their efforts at standardization and mechanization soon paid off, and by 1909, Japan had overtaken China as the

TABLE 15   Major Destinations of Shanghai Raw Silk Exports,
1870–1925 (in piculs)

		Great Britain		France		United States	
	Total	*Weight*	*Percent of Total*	*Weight*	*Percent of Total*	*Weight*	*Percent of Total*
1870	30,482	22,145	72.6	5,639	18.5	1,081	3.5
1875	55,965	20,571	36.8	25,625	45.8	6,132	11.0
1880	69,685	17,620	25.3	36,954	53.0	8,093	11.6
1885	44,690	7,658	17.1	25,938	58.0	6,665	14.9
1890	51,808	9,288	17.9	30,458	58.8	4,714	9.1
1895	76,639	2,463	3.2	41,521	54.2	12,030	15.7
1900	60,432	2,188	3.6	26,134	43.2	10,193	16.9
1905	63,299	1,322	2.1	23,581	37.3	15,544	24.6
1910	87,540	1,148	1.3	34,734	39.7	21,858	25.0
1915	95,822	3,199	3.3	32,821	34.3	41,116	42.9
1920	54,366	1,661	3.1	19,742	36.3	17,874	32.9
1925	120,503	2,213	1.8	51,163	42.5	52,970	44.0

Source: Annual Trade Reports and Returns, annual until 1920. After 1920, *Quarterly Returns of Trade.*

world's leading silk exporter.[60] As Table 16 shows, Japan's raw silk exports increased from 7,000 piculs in 1870 to 549,000 in 1928. Figure 6 shows the dramatic change in China and Japan's relative positions in the export trade.

It was the size and nature of the American market that determined the character and success of the Japanese silk industry. Unlike the French, who still used handlooms for high-quality silks, the American weaving industry in centers such as Paterson, New Jersey used power looms, which required standardized raw materials.[61] It was not that Japanese silk was intrinsically of better quality than Chinese silk, but it was consistent and reliable. According to C. F. Remer, one of the most astute observers of the Chinese economy in the 1920s, "Chinese silk is either excellent or rather poor in quality, and it seems to have been for a long time. Japanese raw silk is of a more uniform quality, but the best Chinese silk is said to be superior to the Japanese product."[62] By the 1920s, Japanese silk accounted for 90 percent of the silk imported into the United States, while Chinese silk accounted for only 10 percent. On the other hand, Chinese silk still retained a relatively more important position in the smaller French market.[63]

Dominated by Japan and China on the supply side, and France and the United States on the demand side, the international silk market nevertheless had elements of instability. While the ultimate source of instability on the supply side was the weather and other natural factors, the whims of taste and fashion were the determinants of instability on the demand side. In the United States, for example, such fashion headlines as the marriage of Alice Roosevelt in a wedding gown of American silk,[64] or the selection of a new range of silk colors for the inaugural gowns of Mrs. Woodrow Wilson and her three daughters, sent the silk manufacturers into ecstasy.[65] The revival of dancing in the United States was hailed as a boon to the silk business.[66] Perhaps the greatest stimulus to the silk industry was the increasing use in the United States of silk hosiery by both men and women, making a virtual necessity out of what had once been only a luxury.[67] Rayon, which had been developed in Europe, did not begin to affect the American consumption of silk very seriously until after 1929. It became a serious competitor to silk, however, in all products in which silk could be used except stockings. Nylon, introduced by Du Pont in 1939, captured the last silk market that had been untouched by rayon.[68]

For the Chinese, the uncertainties of the Western market were exacerbated by the technological risks described in Chapter 1. Moreover, as the export trade grew and accounted for a larger share of total output, the element of uncertainty increased. Until the turn of the century, foreign merchants in China continued to feel that the Chinese domestic market was "incomparably more important" in determining the price of Chinese silk than the foreign market.[69] In 1905, the *Annual Trade Report* of the

TABLE 16 Chinese and Japanese Raw Silk Exports, 1870–1937
(1,000s of piculs)

Year	China	Japan
1870	49	7
1871	60	13
1872	65	9
1873	61	12
1874	75	10
1875	80	12
1876	79	19
1877	59	17
1878	67	15
1879	81	16
1880	82	15
1881	66	18
1882	65	29
1883	65	31
1884	68	21
1885	58	25
1886	77	26
1887	79	31
1888	77	47
1889	93	41
1890	80	21
1891	102	53
1892	101	54
1893	94	37
1894	99	55
1895	111	58
1896	88	39
1897	117	69
1898	109	48
1899	148	59
1900	97	46
1901	129	86

TABLE 16 (continued)

Year	China	Japan
1902	120	81
1903	95	76
1904	125	97
1905	106	72
1906	111	104
1907	116	94
1908	129	115
1909	130	135
1910	139	148
1911	130	145
1912	158	171
1913	149	202
1914	109	171
1915	143	178
1916	122	217
1917	126	258
1918	125	243
1919	166	286
1920	104	175
1921	151	262
1922	143	344
1923	138	263
1924	131	373
1925	168	438
1926	169	443
1927	160	522
1928	180	549
1929	190	575
1930	152	470
1931	136	561
1932	78	549
1933	77	484
1934	54	507

TABLE 16 (continued)

Year	China	Japan
1935	76	555
1936	63	505
1937	69	479

Sources: For China, Table 9.
For Japan, 1870–1914, *Silk: Replies from the Commissioners of Customs,* p. 203.
1915–1930, H. D. Fong, p. 486, Table II.
1931–1937, *Nihon sen'i sangyōshi,* comp. Nihon sen'i kyōgikai (Osaka, 1958), I, 941.

Maritime Customs noted that "native consumption tends to accentuate the deficiency caused by indifferent crops."[70] In the twentieth century, however, prices at Shanghai were clearly determined more at New York, Lyon, and Yokohama than at home.[71]

As a result of domestic and international uncertainties, the silk business was highly speculative, regarded by some as "much more speculative than any other component of China's foreign trade, either import or export."[72] Of the international market, the Silk Association of America wrote in 1911, "Silk is not sold to-day purely on the basis of supply and demand, but has become to a large extent a speculative article. Hence, the trade at large is at a loss how to protect itself against the gambling spirit which has been infused into the trading in this most valuable commodity."[73] One well-known example of speculation was that of the compradore Hu Kuang-yung, who bought and hoarded 15,000 bales of raw silk in 1882 with the expectation that both Chinese and European silkworm crops would be bad the following year. This caused great anxiety among the other silk dealers in Shanghai who feared that he would suddenly flood the market with his stock. Ultimately, when the Italian silk crop turned out to be much better than predicted, Hu was said to have lost one and a half million taels. His bankruptcy had widespread repercussions in the Shanghai financial world, and even on the European market, where his stock was "felt as a dead-weight."[74]

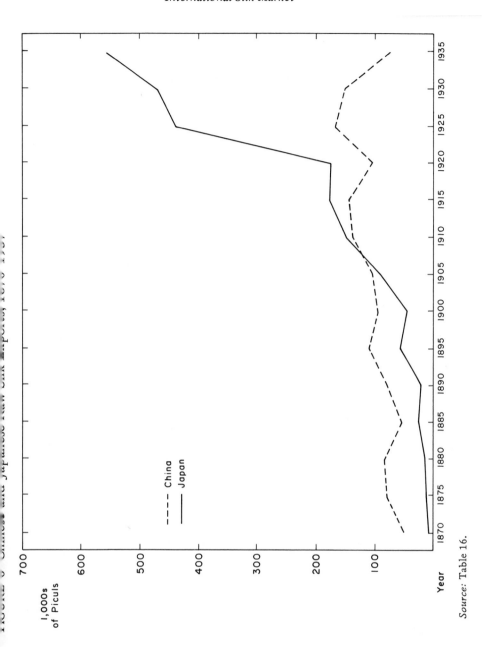

Source: Table 16.

The problem of the exchange rate between gold and silver greatly encouraged another type of speculation. While most of the trading nations of the world gradually shifted to the gold standard, China remained on the silver standard. The long-term decline of the gold value of silver should have served to make Chinese exports more competitive in the world market,[75] but foreign silk merchants complained of a lowering of their profits. But in fact what made a stronger impact on business as a whole were the sudden and dramatic fluctuations in the value of silver rather than the long-term decline. According to Remer, "The result of the rapid fluctuations in silver values . . . was to make the speculative aspect of the trade more important and so to make what may be called illegitimate speculation more common."[76] Since orders had to be placed months in advance, foreign exporters would sell gold for the amount of the order to exchange banks, thus passing the risk of an unfavorable change in exchange rates onto the bank.[77] As steam filatures became well-established in Shanghai, speculation on the annual cocoon crop was common. In 1906, for example, when the Chekiang cocoon crop was down 40 percent, the Customs report observed that "speculation was rife, and prices rose to such an extent that many filatures were unable to cover in full. The rise in exchange caused speculators to lose heavily, and rendered prospects of filatures very unpromising until the last three months of the year."[78]

The Customs authorities at Shanghai believed that business practices had deteriorated since the early days of the treaty port. The opening of the Suez Canal in 1870 and the establishment of direct telegraphic communication between Shanghai and London in 1871 changed the position of the foreign merchant in China.[79] "The dealer was no longer a true 'merchant,' but had become a mere link in the trading chain, bound fast to the interests financing him."[80]

> In the old times business was done in Shanghai by men having command of large capital, who bought heavy consignments here and stored them until there was a chance of sale, or they bought

foreign goods and sent them home to find a market. . . . The Silk trade is largely done now on orders from Europe, and the purchases from the Chinese merchant are not completed until the finance of the transaction is definitely arranged and the laying down calculated to a fraction of a penny.

The daily or hourly fluctuations in exchange may have made this necessary, but . . . it facilitates the carrying on of trade with small or no capital; it promotes a sharpness in business and a keenness in competition which tend to make getting business a more important consideration than how it is got; and it renders it necessary . . . for us Customs people to be far more watchful of frauds than in the days when the Chinese trade was confined to men of more means and making larger profits."[81]

The silk export business at Shanghai was handled almost exclusively by foreign trading firms. In the late 1920s, there were about forty or fifty foreign firms there exporting silk, among which the French and the British were the most numerous. Although there were a few attempts to set up Chinese silk exporting companies in the 1920s, almost all met with failure.[82] Unlike their Japanese counterparts, who by the twentieth century had gotten into the direct exporting of silk and eventually excluded foreign companies from the business, the Chinese silk merchants remained as middlemen between the foreign trading companies and the filature companies and suppliers in the interior. Although this was extremely lucrative for many of them, in the long run it put them at a great disadvantage with respect to their Japanese competitors, whose direct marketing techniques made them very sensitive to changing foreign tastes.

At Shanghai a small group of about fifty silk merchants dominated the export business, serving as the crucial links between producers and foreign exporters.[83] These men served in multiple capacities. First, they often served as compradores of foreign companies, responsible for conducting the firm's business with Chinese producers and dealers. The compradore's relationship to the Chinese dealers was very close; sometimes he lent money to them. Very often he himself was a dealer.[84] In the later years of the silk trade, the function of the compradore as a compradore became less essential, as more Chinese developed competence in Western languages and business procedures.[85]

Second, the Shanghai merchants functioned as independent silk brokers, a role which gradually superseded the compradore's in importance. There were essentially three types of brokers, or *ssu-hao*. The first was the white silk broker, who dealt in hand-reeled silks, primarily Tsatlees. He arranged the financing and storage of Tsatlees for silk hongs in the interior.[86] Only the smaller silk hongs, however, used the services of the brokers; the larger dealers negotiated directly with the foreign exporters.[87] The second type of silk broker dealt mainly with silks produced in Szechwan, Hupei, Shantung, and other provinces. When these silks were shipped to Shanghai for export, they were usually stored in warehouses or godowns until sold or consigned to the foreign exporters.[88] The third type of broker dealt exclusively in filature silk. He represented the filature in the sale of its silk to the exporters, for which service he received a handsome commission.[89] Some brokers dealt exclusively with a certain group of filatures, or a certain group of exporters. Both filatures and exporters preferred to deal through a middleman so that disputes might be settled without direct confrontation.

Third, these merchants often functioned as entrepreneurs. Since a broker, or rather a brokerage, was usually well-financed, it could advance funds to a filature with the understanding that the silk would be marketed through the brokerage. Sometimes one brokerage would have such an arrangement with several filatures.[90] Needless to say, the silk merchants profited very handsomely from their multiple roles. Among the big silk brokers at Shanghai in the 1920s was Hsueh Hao-feng, who was compradore of the French firm of Madier and Ribet, and part-owner of three filatures. There was also Huang Chi-wen, compradore of the American firm of Eagle and Company, and owner of two steam filatures. And there was also Ch'iu Min-t'ing, compradore for the British firm of F. C. Heffer and Company, as well as owner of one filature and part-owner of three others.[91]

The Chinese silk dealers located at Shanghai were organized

along regional lines. From the early treaty port days, when the Cantonese merchants were eclipsed, until about 1911, their main group was the Chekiang, primarily Hu-chou, group. Nan-hsun merchants were supposed to be particularly skilled with foreign languages, which was considered one of the keys to their success.[92] After 1911, however, silk merchants from Wusih became the leaders in the silk business—a reflection of the development of Wusih as a silk filature center.[93] The umbrella organization for the Hu-chou merchants at Shanghai was the Kiangsu and Chekiang Silk and Reeled Silk Merchants' Association, which dealt in Tsatlee silks.[94] The main organization of filature merchants was named the Steam Filature and Cocoon Hong Guild of Kiangsu, Chekiang, and Anhwei. This latter group became the most important silk organization in Shanghai; even foreign companies contributed money towards its support.[95] At Canton the export trade was similarly organized, with commission houses (*ssu-chuang*) serving as intermediaries between the filatures, which accounted for virtually all exports by the 1890s, and the foreign export companies.[96]

The conspicuous failure of Chinese businessmen to get directly into the export business represented the loss of potential profits. Certainly it meant that Chinese manufacturers were less aggressive in their marketing techniques and less alert to the changes of foreign tastes than their Japanese competitors. It was characteristic of the silk business, although not unique to it, that the Shanghai and Canton brokers who served as the middlemen were apparently content with their critical but limited role. After all they were in a position to make large profits without assuming great risks or tying down their capital for very long. More important, the foreign firms had gotten there first, and foreign companies were in firm control of transoceanic shipping, the Japanese having taken up the slack left by the Europeans in World War I.[97]

The organization of the silk export trade at Shanghai, as well as Chinese foreign trade in general, presented a paradox. Although the failure of Chinese merchants to seize control of the exporting business from foreign companies would constitute a serious weakness in the long run, in the nineteenth century, the Western merchants viewed the solidarity of the Chinese mercantile community and the impenetrability of the internal Chinese commercial networks as a source of strength for the Chinese and as a source of frustration and envy for themselves.

> If weak in the arts of organized production, the Chinese had it all their own way in the matter of commercial organization. . . . The foreign merchants . . . were independent and haphazard. They competed against each other, often frantically and uselessly. But the guild organisations had a strong control over the markets . . . [and] there is no doubt that these methods were very successful, passing off losses in bad seasons and reaping full advantage from good and prosperous seasons.[98]

Just as foreigners found it difficult, if not impossible, to control the internal distribution of their imports to China, so too they found it difficult to exercise control over the manufacturing and processing of goods for export.[99] The strong control which Chinese merchants continued to exercise over marketing in the interior as well as sources of supply meant that foreign companies continued, in the broadest sense, to be as dependent on Chinese middlemen as they had been in the days of the Canton trade.

The silk export trade was paradigmatic of Chinese foreign trade in general. Throughout at least the nineteenth century, and even during parts of the twentieth, China regarded foreign trade as being unimportant at best, and harmful to its internal economy at worst. As in earlier periods of history, China felt it had little to gain from importing foreign products, and was content to permit exports if they did not damage the domestic economy. But on the whole, Chinese merchants did not seek to develop Chinese products for foreign markets. Having been forced to concede certain commercial privileges to foreign powers, China tolerated the growth of a foreign trading community in the treaty ports but, at the same time, it did not permit their merchant-agents to expand inward. In the silk trade, foreign companies dominated the export side, without any serious competition from Chinese merchants, but they were not able to exercise any direct control over production and distribution in the domestic sector. Treaty port and hinterland appeared to be, in the words of Rhoads Murphey, "Two Worlds."[100] Appearances, however, are deceiving. They were by no means two hermetically separated worlds. The Western market for silk in the nineteenth century caused an expansion of sericulture in several regions of the interior. The next chapter explores the dimensions and patterns of that growth.

FOUR

Foreign Trade and Domestic Growth

In analyzing the response of Chinese sericulture and silk manufacture to Western trade, there are three dimensions that must be considered: technological, institutional, and quantitative. In the first chapter, I argued that the problem of technological change cannot be understood apart from its institutional context. Neither the technological nor the institutional dimension can be understood, however, without a knowledge of the size, growth, and composition of the industry. Yet it is precisely this quantitative dimension that most eludes our grasp. While there is a full and detailed record of the export trade after 1870, there are no comparable figures for the total output of silk in China, or for the size of the domestic market, even in the twentieth century. Without a sense of the magnitude of the production of silk in China, we have no way of putting the export trade into a proper perspective. If silk exports continued to be a small portion of the total output of Chinese silk, then slowness in technological and institutional change would seem more understandable. If, on the other hand, exports accounted for most of output and largely determined its size and composition, one would expect a greater incentive for technological change, and perhaps institutional changes as well.

Since we have a good picture of exports, if we had some sense of the size of *either* the domestic market *or* total output, we should be able to calculate the other. In this chapter I shall look at estimates for overall output and domestic consumption and then test them against actual evidence concerning, first, the pattern of growth in various regions of China, and second, the development of the silk weaving industry.

The first systematic attempt to study silk production in China was a survey conducted by the Imperial Maritime Customs Service in 1879 and 1880 in response to an inquiry from Natalis Rondot, the representative of the Chamber of Commerce of the city of Lyon, the center of the French silk weaving industry. The Customs Service directed its commissioners to report on the state of sericulture and silk weaving in their districts. The quality of their responses, published in 1881, ranged from perfunctory to painstaking.[1] Rondot himself later published a two-volume work, *Les soies,*[2] which contained estimates of output by province, based on a careful and systematic evaluation of Western and Chinese observations. Rondot believed his figures to be on the conservative side. In the late Ch'ing, the Ministry of Agriculture, Industry, and Commerce published some crop and manufacturing statistics, which attempted prefectural breakdowns and have some usefulness.[3] In the early Republican period, the new Ministry of Agriculture and Commerce attempted a survey of crop production but, since its figures are generally regarded as inaccurate and incomplete, I have not included them in Table 17.[4] In the 1920s and 1930s, however, several foreign groups began systematic surveys of sericulture in China. The Shanghai

TABLE 17 Estimates of Production of Cocoons by Provinces (in piculs)

Province	1880[a]	1915–1917[b]	1925[c]	1926[d]
Chekiang	825,500	876,766	1,000,000	1,140,000
Kiangsu	275,200	266,745	350,000	545,000
Anhwei	10,800	30,000	30,000	97,100
Kwangtung	576,100	768,300	1,000,000	1,057,400
Shantung	24,100	70,000	60,000	110,000
Szechwan	205,800	640,000	600,000	468,000
Hupei	79,100	100,000	100,000	122,900
Hunan	6,500	16,000	20,000	—
Kwangsi	—	12,000	—	55,600
Honan	100,800	—	100,000	42,900
Shansi	—	—	—	6,500
Fukien	—	—	—	3,900
Kiangsi	—	—	—	—
Manchuria	—	—	—	—
Others	17,100	—	70,000	13,000
Total	2,121,000	2,779,811	3,330,000	3,662,300

Notes: [a]Natalis Rondot, *Les soies* (Paris, 1885–1887), I, 194–208. These figures are converted from Rondot's figures for kilograms of raw silk, using the conversion formula 13 parts fresh cocoons = 1 part raw silk; and 1 picul = 60 kilograms. By using the ratio 15 parts cocoons to 1 part raw silk, Tseng T'ung-ch'un, who reproduces these figures on pp. 79–81, arrives at the total of 2,440,000 piculs of cocoons, but I believe that the results derived from using the 13:1 formula are more consistent with late nineteenth-century standards. See Table 3. If we apply a 10:1, or 11:1, standard, the output figures would be significantly lower.

[b]Akashi Hiroshi, 1918, as cited in Yueh Ssu-ping, p. 13, and Tseng, pp. 74–78. Akashi was apparently deputed by the Japanese Ministry of Agriculture and Commerce to survey the silk scene in China. Tseng reproduces Akashi's explanation for his estimates. In each case his estimate was based on the known supply for the steam filatures, estimates of hand-reeled silk based on known export figures, estimates of the domestic consumption of silk, and other known data for 1915–1917. Akashi's total of 2,779,811 piculs is incomplete, since it excludes data from Honan and other provinces.

[c]*A Survey of the Silk Industry of Central China*, pp. 4 and 96. No explanation of how these figures were derived is given.

[d]Uehara Shigemi was the chief author of *Shina sanshigyō taikan*. According to the prefaces to this work, Uehara studied at the Tōa dōbunkai's school at Shanghai and spent about 5 years traveling through 18 provinces of China. His survey results are the most complete and reliable of all the compendia dealing with sericulture in China. The data reproduced here are found on pp. 10–16, together with a full explanation of how Uehara derived his estimates. This material also appears in Yueh Ssu-ping, p. 14.

International Testing House published a very solid study of silk production in central China in 1925, and several Japanese individuals and groups conducted rural surveys, which constitute probably the most accurate sources on all aspects of silk production in China.

With the wider use of steam filatures, output was easier to estimate because the supply and distribution of cocoons became centralized. Even so, there was no way to estimate accurately the output of cocoons that were still reeled domestically. Moreover, cocoon output estimates provide only an approximate guide to the amount of raw silk actually produced, since the yields of silk from cocoons could vary greatly according to their quality and whether they had been dried or not. (See Chapter 1, Table 3.) For example, the 1880 cocoon output estimates in Table 17 would have been significantly lower if a 10:1 ratio had been used in converting them from silk output figures. Estimates of mulberry output were even less reliable than those for cocoons or silk, because mulberry was often grown in marginal areas rather than in discrete fields which could be measured. Added to these difficulties was the general problem of lack of uniform weights and measures in China. At Wu-ch'ing, for example, a picul of leaves could weigh between seventy and ninety catties.[5]

With these qualifications in mind, let us examine some of the estimates of cocoon output from 1880 to 1926, as shown in Table 17. These figures show that total output rose from 2,121,000 piculs in 1880 to 3,662,300 in 1926, an increase of 72.7 percent. In this same period, the total exports of raw silk and silk fabrics grew from the equivalent of 1,170,000 piculs of cocoons to the equivalent of 2,574,000 piculs of cocoons, an increase of 120 percent, as shown in Table 18. In short, exports expanded at a faster rate than total output.

The provincial output figures in Table 17 show that the Kiang-nan provinces of Chekiang and Kiangsu, together with Anhwei, produced more than half of the national output at all times. Kwangtung province's share of national output increased from

TABLE 18 Estimates of the Growth and Composition
of Cocoon Production, 1880–1926
(1,000s of piculs)

Date	(a) Total Output of Silk	(b) Raw Silk Exports	(c) Silk Fabric Exports	(d) Total Exports (b + c) or (a −e)	(e) Domestic Consumption (a − d) or (f −c)	(f) Domestic Weaving (c + e) or (a −b)
1880	2,121	1,066	104	1,170	951	1,055
1917	2,929	1,638	390	2,028	901	1,291
1925	3,330	2,184	403	2,587	743	1,146
1926	3,662	2,197	377	2,574	1,088	1,465

Sources: [a]Table 17. I have added 150,000 piculs to the 1915–1917 figure, which did not include Honan and miscellaneous provinces.
[b]Table 9. Converted from piculs of raw silk at 13:1.
[c]Ibid.

about one-quarter to one-third during these years. Anhwei and Shantung provinces showed the fastest rate of growth. In the former, sericulture was stimulated by the development of Shanghai silk filatures, while in the latter, sudden growth was due to the popularity of wild or Tussah silk. Manchuria was also affected by the popularity of Tussahs, and was presumably included in the "Other" category of the surveys used in compiling Table 17. In three provinces—Kiangsu, Kwangtung, and Szechwan—the output doubled, primarily in response to foreign trade in the first two cases. Of the major cocoon-producing provinces, the most important, Chekiang, showed the slowest rate of growth, only about 38 percent.

If we take the estimates of total output and subtract from them the figures for total silk exports, we have an estimate of domestic consumption. As shown in Table 18 and Figure 7, domestic consumption increased very slightly throughout this period, from 951,000 piculs of cocoons in 1880 to 1,088,000 piculs in 1926. Thus, the share of total output represented by domestic consumption declined from about 44.8 percent to 29.7 percent, according to this set of estimates. Most of the

FIGURE 7   Estimates of Growth and Composition of
Cocoon Production, 1880–1926

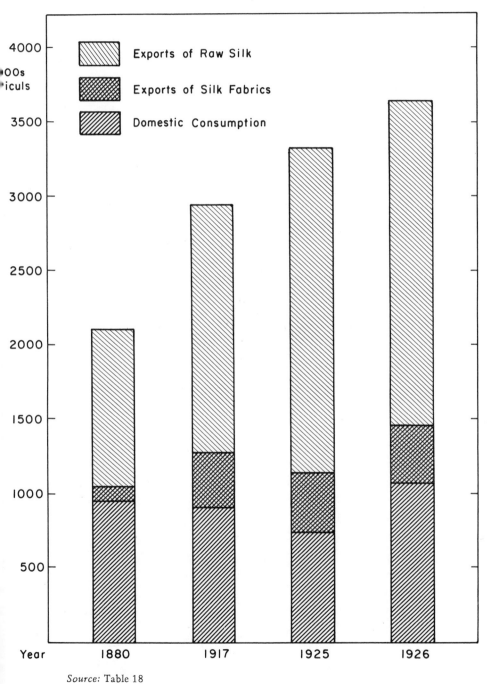

Source: Table 18

studies done in the 1920s and 1930s, however, claimed that the level of domestic consumption was significantly higher. Japanese experts estimated domestic consumption at 45 percent of total output.[6] A survey of Canton's silk industry published by Lingnan College in 1925 suggested that more then 60 percent was domestically consumed.[7] And H. D. Fong of the Nankai Institute estimated that 55 percent was retained for domestic use in the early 1930s.[8] The weight of all this professional opinion suggests that the output figures in Table 17 may be underestimated by a significant degree.

Nevertheless, these estimates provide some insight into the fate of the Chinese silk weaving industry in the modern era. As Table 18 and Figure 7 show, silk fabric exports expanded in both relative and absolute terms from the late nineteenth to the early twentieth century, from which time onward they held a steady 17–24 percent share of total silk exports. These fabric export figures added to the domestic consumption figures give us the total output of silk fabrics in China. Since this output remained relatively steady during these years, and piece goods continued to play an important role in total exports, it appears that after recovery from the Taiping Rebellion, the silk weaving industry experienced neither net expansion nor contraction as a result of the export trade in raw silk.

These estimates and findings suggest the following scenario for the development of the silk industry in China between 1842 and 1937: the opening of the treaty ports coincided with a decline in the domestic demand for raw silk primarily because of decreased production at the Imperial Silk Factories. As foreigners observed, the amount of silk exported at this time was still small when compared to total output, and therefore no particular dislocation was experienced. During the Taiping Rebellion, the urban weaving industry at Soochow, Hangchow, and Nanking suffered serious damage, from which recovery was slow and incomplete. Although the sericultural areas in the countryside also experienced some devastation, they could still produce enough raw silk during the 1860s and 1870s to meet the growing foreign demand. By the 1880s, the weaving industry had largely recovered, although not to its eighteenth-century proportions, and maintained a fairly steady foreign and domestic market until the 1930s. There was still a considerable demand for silk and silk products in China, but it did not increase proportionately with the treaty port population primarily because fine cottons and, later, artificial fibers competed for the domestic market for luxury fabrics. From the 1880s on, further increases in silk exports could no longer be made at the expense of the domestic sector, and, instead, resulted in net expansion of silk production both in Kiangnan and in other regions of China. The pattern of provincial development suggests that foreign trade stimulated the most rapid growth in those areas newly introduced to sericulture, and that its impact, in quantitative terms at least, was relatively muted in the more developed, traditional sericultural areas. Some new areas tended to be more responsive to technological innovation than the old areas.

GROWTH IN KIANGNAN

Since the Kiangnan area was the single most important silk-growing region of China, by examining in depth its pattern of development after 1842, it may be possible to test the general propositions about the rate and shape of growth that have just been suggested. Although the region had a certain historical and geographical unity, with respect to sericulture, localities within it had different experiences. It is important to distinguish between the old sericultural districts south of T'ai-hu, primarily in Hu-chou and Chia-hsing prefectures, which had traditionally been the source of the best raw silk for the Imperial Silk Weaving Factories, and the new sericultural areas, which developed in the late nineteenth century mainly in response to trade with the West (see Map 1).

In Hu-chou and Chia-hsing, the opening of Shanghai as a

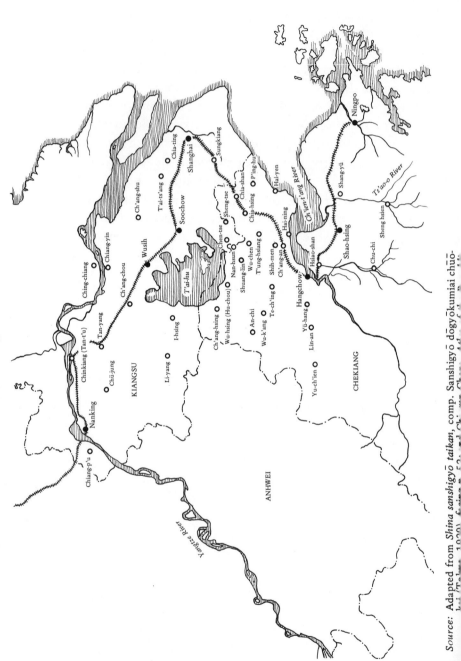

MAP 1   Silk-Producing Localities in Kiangnan (Republican Period)

Source: Adapted from *Shina sanshigyō taikan*, comp. Sanshigyō dōgyōkumiai chuō-

treaty port meant that production and trade were partially redirected through Shanghai to new foreign markets. During the 1850s, much of the silk exported was probably made available by the disruption in the normal domestic markets caused by the Taiping Rebellion.[9] The rebellion itself did not directly affect the export trade until 1860, when the Taiping forces began to occupy the area around Shanghai. Nan-hsun, the most important silk center in Hu-chou, was occupied from January 1862 to August 1864. Although the occupation did much damage to neighboring villages and to mulberry trees,[10] local merchants somehow arranged for the safe passage of boats transporting silk to Shanghai.[11] Some local gazetteers said that profits and taxes from the silk trade contributed a significant share of the cost of feeding the anti-Taiping forces and of reconstruction later.[12]

The effect of the rebellion seemed to vary widely even within these two prefectures. One commissioner of customs reported from Chia-hsing prefectural city in 1878, "I found it almost in the same state as when I saw it twelve years ago. Very few houses have been rebuilt, a great portion of the city is still in ruins, and the suburbs, which before their occupation by the rebels were so important, have entirely disappeared."[13] In some localities in Hu-chou, the population had been so greatly reduced by the rebellion that local authorities developed a policy of welcoming new settlers from other provinces.[14] For the most part, however, recovery in the rural areas was quite rapid, and the restoration of the silk districts was complete by 1877.[15] The Maritime Customs commissioners reported that "as far as could be ascertained on inquiry in the Silk Districts, the country people and manufacturers agree in saying that the Taiping insurrection did not do any harm to the quality of silk manufactures. It is, however, well known that a great number of mulberry trees having been cut down or rooted up by the rebels, the production of Silk decreased for a time very considerably."[16] At Nan-hsun, one commissioner reported that "though this district was long in the hands of the rebels, the inhabitants are so active and the trade so prosperous that all traces of the rebel occupation have entirely disappeared."[17]

After recovery, the old silk districts intensified and expanded their cultivation of mulberry trees and silkworms, and tremendous prosperity resulted. At Nan-hsun, business boomed; it was said that in a good year people could earn a whole year's income for a month's work during the silk season.[18] Ling-hu, Kuei-an, and Te-ch'ing were other localities in Hu-chou prefecture which reaped great benefits from the silk trade.[19] In An-chi hsien it was reported that "in all the *hsiang* and *chou,* the people are raising silkworms, and the mulberry trees are flourishing. Recently mulberry trees have been planted even in the hills."[20] In many places sericulture was undertaken for the first time. For example, at P'ing-hu in Chia-hsing prefecture, "in the past, only the southwest *hsiang* had mulberry trees, but now to the east of the city, for 20–30 *li* by water, there is absolutely no unclaimed land left, and the peasants do [sericulture] as their regular occupation."[21] In Hai-yen hsien, sericulture had been practiced since the late Ming, but after the Taiping Rebellion, it was reported that "mulberry and *che* trees grow everywhere, in the fields and along the walls; sometimes they are planted with wheat. There is not one person who does not practice sericulture. The techniques have all been learned from Wu-hsing."[22] At Shih-men hsien, the local gazetteer observed that although the gentry and scholars were all learned in literature and philosophy, their real preoccupation was the pursuit of profit in the silk business.[23]

In Hu-chou and Chia-hsing, cocoons were reeled at home for export as Tsatlees and Taysaams. Even in the 1920s, after the establishment of filatures at Shanghai, most of the silk produced in Chekiang continued to be domestically reeled, as shown in Table 19. At Chia-hsing a few families sold their cocoons for filature reeling, but most continued to reel their own silk.[24] One exception, however, was found at P'ing-hu, where very few peasants after the establishment of seven filatures in the area continued to do their own reeling. The people

TABLE 19 Estimate of Fresh Cocoons Used in Domestic
and Filature Silk in Kiangnan, c. 1926
(in piculs)

Province	Domestic	Filature	Total	Filature as Percentage of Total
Chekiang	900,000	240,000	1,140,000	21.1
Kiangsu	45,500	499,500	545,000	91.7
Anhwei	22,100	75,000	97,100	77.2
Total	967,600	814,500	1,782,100	45.7

Source: Shina sanshigyō taikan pp. 10–12. See also *A Survey of the Silk Industry of Central China*, pp. 5–6, and 93, for comparable breakdowns.

of P'ing-hu were considered unusually willing to adopt new methods.[25] On the whole, however, since there remained both a domestic and foreign market for Tsatlees and Taysaams, there was no pressing need for Hu-chou and Chia-hsing to change their method of production.

The collapse of the world market for silk after the Great Depression affected these old silk areas the most severely since they were the most highly specialized. The amount of land devoted to mulberry cultivation declined quickly.[26] In 1930 there were ninety-four cocoon hongs in Wu-hsing, but by 1932, only one was still in business.[27] In nearby Ch'ung-te, the income of families surveyed dropped to 57 percent of their previous average income.[28] Not only was there a loss of income from sericulture in these areas but, since they were dependent on rice imports from other regions, they were doubly vulnerable to economic shifts.[29]

In Shao-hsing and Hangchow prefectures, the silk export trade did not have a pronounced effect until the twentieth century. In Shao-hsing sericulture had never been very important.[30] Before the Kuang-hsu period, only Chu-chi hsien had produced much silk.[31] In the twentieth century, however, Shao-hsing accounted for about 20 percent of all cocoons produced in Chekiang, as shown in Table 20. The Ts'ao-o River valley, along which sericulture was concentrated, was the only major silk district in Chekiang that produced cocoons for the filatures rather than reeling its own silk.[32] In Hangchow prefecture, some sericulture had been practiced traditionally, with Hai-ning and Jen-ho hsiens producing the most silk, and Lin-an having a particularly good reputation for reeling.[33] After the Taiping Rebellion, silk output increased gradually, but served to supply the needs of local weavers rather than the foreign market.[34] In the twentieth century, Hai-ning became a particularly important producer for the export trade. By the 1920s, mulberry fields were said to represent 45 percent of all cultivated land in some localities in this area. Most of the cocoons were still reeled by hand.[35]

By contrast, in Kiangsu province, sericultural activity was overwhelmingly directed toward supplying steam filatures, particularly after Wusih became a steam filature center. Kiangsu showed much more dramatic growth than Chekiang, primarily because sericulture was a relatively new development. Prior to Western trade, silk had been produced only in small quantities to supplement the supply of raw silk from Hu-chou and Chia-hsing for local weavers. The process of change was most pronounced in the three southwestern prefectures of Kiangsu: Ch'ang-chou, Chinkiang, and Chiang-ning. Since these three prefectures had experienced particularly heavy destruction during the Taiping Rebellion, they emerged from the rebellion as a severely depopulated wasteland, into which people from Chekiang, north Kiangsu, Anhwei, Hupei, and Honan migrated after the rebellion. Because of the vast tracts of land which had been left uncultivated during the rebellion, it was easier to introduce new crops afterwards. Of these, mulberry was the most popular, and spread all over these three prefectures.[36]

In Chiang-ning prefecture, the recovery of sericulture was relatively rapid despite the severe devastation caused by the Taipings at Nanking.[37] Virtually all of the raw silk produced in this region was for the use of the local silk weaving industry rather than for export.[38] Much of the revival was sponsored by local officials, including Tso Tsung-t'ang, who, as governor-general of Liang-Kiang in 1882, ordered mulberry saplings be distributed at Chiang-p'u and Chü-jung, where it is said they brought "leisure and profit" to the people.[39] The process of official promotion of sericulture continued into the twentieth century. In 1905, deputies of the provincial treasurer went to Chekiang to buy 150,000 mulberry saplings, which were distributed to surrounding localities.[40] Still, on the whole, the output of silk from this prefecture was relatively small.

In Chinkiang prefecture, the same pattern of official promotion was repeated, but recovery from the Taiping Rebellion was slow.[41] At Tan-yang, where little silk had been produced before the rebellion, local officials taught the people how to plant and cultivate the *Hu-sang* in their devastated fields. It was said that the profits were considerable.[42] In Tan-t'u hsien, the local taotai, Shen Ping-ch'eng established a mulberry bureau to teach people how to plant *Hu-sang*.[43] Of all the Chinkiang districts, Li-yang was the most highly developed in sericulture, and a great proportion of its peasant families depended on it as their main source of income.[44] It had a tradition of sericulture dating back at least to the Ch'ien-lung period, but after the rebellion, most of its silk was sent to Shanghai for export.[45]

It was in Ch'ang-chou prefecture where the greatest growth in sericulture took place. Wusih and its twin city, Chin-kuei, had never produced much silk before the rebellion, but afterwards sericulture quickly became popular.[46] By 1880 Wusih was producing about 2,000 piculs of raw silk a year, of which about 40 percent was sent to Shanghai for export and the rest used in local weaving.[47] Wusih's real growth, however, occurred in the twentieth century, when it became a leading center for the production and sale of cocoons for steam filatures. When the northern Chekiang crop became insufficient for the needs of the Shanghai steam filatures, buyers started turning toward Wusih.[48] With the completion of the Shanghai-Nanking Railway in 1908, it became feasible to transport cocoons to

TABLE 20 Estimates of Cocoon Production in Chekiang[a]
(in piculs)

Place	1880[b]	1910[c]	1933[d]
HU-CHOU FU	363,461	312,701	282,900
Kuei-an	—	115,236	—
Wu-hsing			203,400
Wu-ch'eng	—	116,089	—
Ch'ang-hsing	—	32,935	45,200
An-chi	—	2,602	3,380
Hsiao-feng	—	3,289	3,800
Wu-k'ang	—	5,386	7,120
Te-ch'ing	—	37,164	20,000
CHIA-HSING FU	93,903	136,423	319,500
Chia-hsing	—	29,358	175,200
T'ung-hsiang	—	21,645	54,300
Shih-men	—	27,893	20,000
(Ch'ung-te)			
Hai-yen	—	32,867	41,800
Hsiu-shui	—	14,283	—
P'ing-hu	—	6,732	25,200
Chia-shan	—	3,645	3,000
HANGCHOW FU	102,875	104,819	379,300
Ch'ien-t'ang	—	29,896	—
Hangchow			189,000
Jen-ho	—	21,628	—
Yü-hang	—	3,312	34,900
Lin-an	—	943	24,300
Yü-ch'ien	—	1,289	3,800
Ch'ang-hua	—	589	1,700
Hsin-ch'eng	—	2,512	6,600
Fu-yang	—	4,754	9,000
Hai-ning chou	—	39,896	110,000

TABLE 20 (continued)

Place	1880[b]	1910[c]	1933[d]
SHAO-HSING FU	21,497	—	270,090
Hsiao-shan	—	—	129,660
Chu-chi	—	—	54,000
Hsin-ch'ang	—	—	18,200
Yü-yao	—	—	—
Shang-yü	—	—	3,680
K'uai-chi ⎫	—	—	6,500
⎬ Shao-hsing			
Shan-yin ⎭	—	—	—
Sheng	—	—	56,950
Other	—	—	1,100
TOTAL — For Four			
Prefectures	581,736	—	1,251,790
For Three			
Prefectures	—	553,943	—

Notes: [a]In this table and the following ones, prefectural divisions have been re-tained for the sake of organization, although the prefectures were abolished after 1911. Most of the hsien remained the same after 1911, except the name of Shih-men hsien was changed to Ch'ung-te; Kuei-an and Wu-ch'eng hsiens were merged to form Wu-hsing hsien; Ch'ien-t'ang and Jen-ho became Hangchow hsien; and K'uai-chi and Shan-yin merged to form Shao-hsing hsien. Only four prefectures of Chekiang are listed in these tables, because the size of sericultural undertakings in the other prefectures was relatively unimportant.
[b]*Silk*, pp. 76, 80, 81, and 82. These tables give estimates of the output of raw silk in Chekiang prefectures. To convert these figures to cocoon weights, I have used the 11:1 ratio preferred by this study, p. 53. The results from the tables on these pages are slightly lower than another set of estimates found on p. 111. The figures on pp. 76–82 are further broken down into hsien and *chou* estimates, but I have omitted these breakdowns because the administrative divisions used are not the conventional ones, and there is no way to show comparability with figures from later years.
[c]*Shina seisan jigyō tōkeihyō*, I, 1067ff. These statistics on crops were compiled by the Ministry of Agriculture, Industry, and Commerce of the Ch'ing government in its last years, and reprinted in 1912 by a Japanese group in Tientsin.
[d]*Chung-kuo shih-yeh chih: Che-chiang sheng*, comp. Chung-kuo shih-yeh pu, (Shanghai, 1933), IV, 184–185. Although reprinted in 1933, the data is probably from a 1920s survey. No source is given.

Shanghai for export. Peasants who had previously raised silk-worms only as a side-occupation began to expand their scale of production and the price of cocoons began to rise. Many of the mulberry trees seen in the area from Wusih to Ch'ang-chou were planted after 1911.[49] Mulberry fields occupied about 30 percent of the total cultivated acreage and accounted for about 70 percent of the value of agricultural output by 1930. According to one estimate, 145,000–200,000 piculs of fresh cocoons were produced annually.[50] By the 1920s, Wusih was exclusively a cocoon-producing center, with little reeling done at home. In fact, Wusih became an importer of cocoons, since the needs of the local filatures far exceeded the local supply of cocoons. By 1925 there were at least eighteen filatures and, according to one estimate, only one-third of their supply came from local sources.[51]

Finally, in Anhwei province, the development of sericulture was also comparatively recent and related to the rise of steam filatures in Shanghai. The bulk of the cocoon output of the province was shipped to Shanghai for reeling, although its quality was generally considered very poor.[52] Only 22,100 piculs of the province's total output of 97,100 piculs in 1926 were not machine-reeled, and part of that was waste silk.[53]

This survey of the growth of sericulture in the Kiangnan region roughly confirms the hypothesis that sericulture developed most rapidly in new areas, and that the new areas were more closely tied into the development of the steam filature industry. Certainly the silk-growing regions of Kiangsu and Anhwei followed this pattern. In Chekiang, Hangchow experienced rapid growth without being tied to the modernized sector of the export trade, providing a single exception to the pattern. As Table 19 shows, the estimates of output made by Japanese experts in the 1910s and 1920s confirms the view of a clear separation between two sectors, one oriented toward the traditional reeling and weaving; the other oriented toward the modernized reeling industry.

Did the other major silk-producing regions in China conform to this pattern of development? A brief but systematic look at them will test our working hypothesis.

GROWTH IN OTHER REGIONS

Kwangtung province (See Map 2), the second most important silk region, was a relative newcomer to silk production. Sericulture had originated in the late Ming or early Ch'ing in response to the growing Canton trade in silk. Shun-te hsien, the most productive district, developed sericulture in the eighteenth century, and by the mid-nineteenth century was exporting about 10,000 bales, or 7,000 piculs of silk a year to the West.[54] In addition, there were other silk exports from Canton which went to Macao and Hong Kong by junk and were not recorded in the early Customs records. In the earliest days of the treaty ports, Canton also had a larger piece-goods export trade than Shanghai, which initially played a more important role in Canton's exports than raw silk.[55] By 1879, about 21,000–22,000 piculs of raw silk were exported—mostly to Europe and the United States, but some to Bombay. In addition, about 5,000 piculs of piece goods were shipped to Southeast Asia via Hong Kong. According to the Customs estimate, this left about 20,000 piculs of silk for domestic consumption.[56] By the end of the century, however, the raw silk export trade had damaged the piece-goods trade by raising the value of cocoons so that local weavers were less able to compete.[57] The weaving industry at Fo-shan, once the largest in the delta, sharply declined in the early Republican period.[58] By the 1920s, according to some estimates, about 70 percent of raw silk output was exported.[59] If, as the estimates in Table 17 suggest, total output had doubled in the same period, then the extent to which growth was stimulated by foreign trade can be seen.

Like Kiangsu province and unlike Chekiang, sericulture in Kwangtung became almost completely linked to the development of the steam filature industry. Steam filatures began to dominate the export trade as soon as they were established, and by the end of the century they accounted for as much

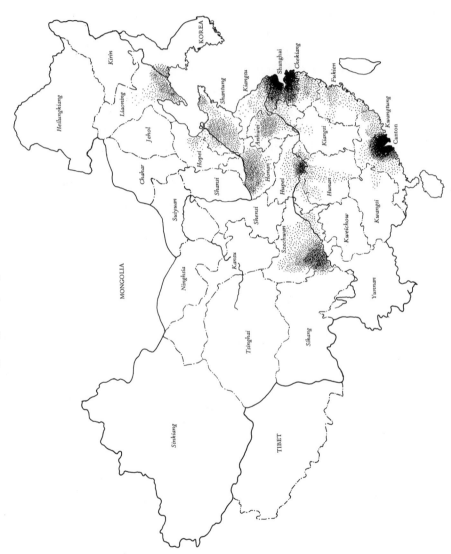

Map 2 China: Silk-Producing Regions (Republican Period)

as 93.5 percent of all raw silk exports.[60] These facts serve to confirm the idea that the growth of sericulture was more pronounced in relatively new areas, and that new areas were relatively more receptive to technological modernization than the older silk areas. It should be noted, however, that Kwangtung was unique among all the silk-producing regions of China in that it was able to support six or seven crops of worms and leaves per year, and many peasant households who raised silkworms did so as a full-time occupation rather than a subsidiary handicraft.[61] Thus the development of the filature industry in Canton was facilitated not only by the relative novelty of sericulture, but also by the fact that it could be supported by year-round and almost full-time sericultural activity in the countryside.

Kwangtung was also unique in the extent to which the economy of its silk districts, centered in the Canton delta, was almost exclusively dependent on sericulture. With the expansion of silk production in the nineteenth century, the very landscape of Shun-te, Nan-hai, Hsiang-shan (renamed Chung-shan after 1927), and other districts in the delta was transformed. Often fish ponds were dug, while the embankments around them were built up and planted with mulberry. Thus sericulture and fishing became two important and interdependent occupations.[62] By the 1920s, approximately 70 percent of Shun-te's land, or about 665,000 mou, was devoted to mulberry fields, and it was estimated that 1,440,000 people, out of a total population of 1,800,000 people, were involved in some aspect of sericulture. In Hsiang-shan district about 328,000 mou were devoted to mulberry, and a population of about 382,000 was engaged in sericulture. At Nan-hai, the figures were approximately 300,000 mou and 200,000 people. This was a big business and was responsible for producing about one-seventh of the world output of silk.[63]

Szechwan province, like Kiangsu and Kwangtung, was a region where the output of cocoons probably doubled between 1879 and 1926. Throughout this period, however, its silk industry remained relatively unmodernized, and its output of silk was primarily directed toward domestic and non-Western foreign markets, rather than to the Western markets. Almost all Szechwan silk was yellow silk, which was very strong and often used for industrial or decorative purposes, but not for weaving high-quality silk fabrics.[64] Therefore, Szechwan was not involved in direct competition with Shanghai, Canton, or Yokohama for the international white silk market. Since most of its silk was produced for the domestic market, and since transportation from Szechwan to the coastal ports was costly, there was little incentive for modernization. Filatures were not established until about 1905, and even then they were not equipped with modern machinery, but with wooden hand-machines. Although steam filatures became more important in the Republican period, they were not as numerous as factories with wooden hand-reels.[65]

Customs figures show that steam filature yellow silk did not really become an important export commodity until 1925, and ranged between 10,000–18,000 piculs annually in the next few years.[66] Around 1926, according to another source, of a total of 35,000 piculs of raw silk output from Szechwan, only 6,500 were machine-reeled, almost all for export. Of the remaining 28,500 piculs of hand-reeled silk, 8,500 were exported and 20,000 piculs were retained for domestic consumption. In short, at least 57 percent of raw silk output was used in local weaving.[67] Much of the output of the local weaving industry, which was well established at Ch'eng-tu and Chia-ting, was directed toward non-Western foreign markets, such as India.[68] For these markets, as well as the Chinese market, the rereeling of silks, using Japanese techniques, became widespread in Szechwan.[69] Hand-reeling, as well as rereeling, was organized in Szechwan in workshops as well as in households. Thus, silk production there occupied an intermediate stage between domestic handicrafts and mechanized factory production. While its technology remained basically traditional, Szechwan's silk industry was innovative within that context. In short, although Szechwan's output expanded dramatically in the late nineteenth and early twentieth centuries, the dynamics of this growth were unlike those of either the central or south Chinese regions.

Hupei was a province that closely resembled Szechwan in several respects. First, it was an area where sericulture had traditionally played an important role, and then experienced a period of decline followed by a distinct period of relatively modern growth, most notably in the Han River valley, an area of eighteenth-century settlement.[70] Second, it produced mostly yellow silk of a rather inferior quality.[71] If anything, it was more backward in its reeling techniques than Szechwan. The relatively low unit price of yellow silk probably did not justify the investment in steam filatures. Moreover, Hupei was even less involved in the Western export trade than Szechwan, exporting about 10,000 bales of silk annually mostly to India and the Middle East.[72]

Although its total output was relatively small, Shantung showed the fastest rate of sericultural growth during the period under consideration, according to the estimates in Table 17. Like Szechwan and Hupei, it had been an important sericultural area during the earlier dynasties, but it had not been a major producer during the Ming and Ch'ing periods. Its modern revival was due entirely to the sudden international popularity of Tussahs and Pongees. Exports of Tussahs from China as a whole reached about 30,000 piculs a year during the Republican period, while exports of Pongee silks ranged between 11,000 and 27,000 piculs a year.[73] Manchuria and Shantung were the main sources of wild silk, the former supplying about 70 percent of all wild silk exports from China.[74] Much of the Manchurian output of cocoons was reeled at Chefoo until about 1915, when steam filatures were established in Manchuria, primarily at Antung. Chefoo also developed its own steam filature industry after 1915, but it suffered by not being able to secure an adequate supply of cocoons from Manchuria.[75] This modern reeling industry, financed almost completely by Japanese and other foreign capital, was clearly distinguished from the traditional reeling sector, which still reeled yellow silk, also a local product.[76] Around 1926, at least 45,000 piculs of Shantung's annual output of 110,000 piculs of fresh cocoons was still reeled domestically.[77]

This review of the major silk-growing provinces permits us to make the following general statements: first, there was a significant expansion of silk output between 1879 and 1937 due to foreign trade. Second, this expansion was more pronounced in Kiangsu, Anhwei, Kwangtung, Szechwan, and Shantung—provinces which had not been the major silk producers in the Ming and early Ch'ing—than it was in Chekiang, which had been the major supplier of silk. In almost all cases, except that of Szechwan, the new areas of development were linked to the modern steam filature industry, although the degree and success of mechanization was determined by other factors, such as Kwangtung's favorable climate which permitted year-round production, or Szechwan's distance from the coastal ports, which hindered industrialization. In most cases—Szechwan, Shantung, and Kiangsu—the modernized sector was directed toward the Western market, while the traditional sector continued to produce for the domestic market. Only in Kwangtung did the modern export sector virtually dominate sericultural production. And only in Chekiang did the traditional sector produce a significant amount of domestically reeled silk for the export trade as opposed to the domestic market.

THE SILK WEAVING INDUSTRY AND THE DOMESTIC MARKET

The pattern of sericultural development suggested in this chapter rests on certain assumptions about the domestic market for Chinese silk. Although there is no way to measure directly the size of that market, it is at least possible to see if those assumptions are consistent with available evidence concerning the nature and structure of the silk weaving industry. Is the picture suggested in Figure 7 an accurate one—namely, that after 1880 the silk weaving industry maintained a steady output of silk fabric, of which one-quarter to one-third was exported? To what extent was the growth of the industry hindered by the raw silk export trade, which competed for the same raw materials? Conversely, to what extent was the growth of the export trade restricted by the existence of a strong weaving industry? In answering these questions, we shall focus on the Kiangnan region, which was not only the main sericultural area in China, but also its main silk weaving center.

In Kiangnan it is important to distinguish the three major weaving centers at Nanking, Soochow, and Hangchow, where weaving of luxury silks was done in workshops, from the smaller weaving centers where weaving of lower quality silks was still done domestically. During the nineteenth century, the three centers suffered two blows from which they never fully recovered. The first was the decline in the Imperial demand for luxury silks, which had been the main market for their silks. The second was the Taiping Rebellion, the impact of which was much more direct and severe in these urban centers than on the rural sericultural districts. Since the urban weaving industry used only the best quality white silk from the Hu-chou area, its decline permitted a portion of the raw silk output to be diverted to the export trade during the mid-nineteenth century.

As Table 21 shows, Nanking, which fared the worst, probably had at its peak close to 50,000 looms, including both those inside and outside the city. Of this number, at least 30,000 were satin looms.[78] On the eve of the Taiping occupation in 1853, however, the number of looms had already declined to 14,000–18,000.[79] When the Taiping forces entered Nanking, local weavers fled north to other provinces and southeast to Soochow, Sungkiang, Shanghai, and other localities, which benefited from the skills of these migrants.[80] Recovery after the rebellion was slow and difficult. In order to encourage weavers to return to Nanking and resume their work, the Governor-General Tseng Kuo-fan, ordered a remission of likin taxes on silk goods, established a reconstruction bureau,

TABLE 21   Estimates of Number of Looms in Kiangnan,
                 1880–1930s

Location	Before Taiping Rebellion	1880	1901–1911	1912–1926	1930s
Nanking	14–50,000	5–10,000	6–13,000	10,000	1,000
Soochow	12,000	5,500	9–10,000	9,000	2,500
Hangchow	–	3,000	4,300	7,000	1,000
Sheng-tse	–	8,000	–	–	4,000
Hu-chou	–	4,000	–	13,000	4,000
Shanghai	–	–	–	–	6,000

Source: See text and notes to Chapter 4.

and also a sericulture bureau.[81] Despite these efforts, by 1880 the industry had experienced only a modest recovery from the rebellion. There were about 5,000 satin looms in the city and its immediate suburbs, and another 3,000–6,000 looms which wove silk ribbons and other piece goods. At the turn of the century, the numbers were still roughly the same.[82]

   Both the Boxer Rebellion and the 1911 Revolution had an adverse effect on Nanking and the other centers by cutting off their northern markets. After the Boxer Rebellion, when many Nanking weavers were forced to stop work, Governor-General Liu K'un-i proposed that the wealthy silk merchants contribute money toward the establishment of a government bureau which would buy up silk goods for which there was no immediate market, but apparently his effort had no effect.[83] After a decade of partial recovery, with probably 6,000 or 7,000 looms in operation, Nanking again lost some of its north China markets, as well as those in Mongolia and Tibet, during the 1911 Revolution.[84] During the 1910s, Nanking was reported to have about 10,000 operating looms, which produced 200,000 bolts of silk a year, and kept 50,000 weavers employed. It is estimated that at least 200,000 people in Nanking depended on the silk industry in one way or another for their livelihood.[85]

In the long run, however, Nanking was unable to compete with the other weaving centers. By the 1930s, there were only about a thousand looms inside the city and five hundred outside, producing only 23,000 bolts of silk.[86] Thousands who had previously depended on weaving were forced to find new employment.[87] Nanking's weaving industry was hopelessly backward, with no modern equipment at all, and outdated patterns and techniques, for which there was little demand.[88] One reason why Nanking experienced greater difficulty in the modern period than Soochow or Hangchow was its relative distance from the main sericultural areas of Hu-chou and Chia-hsing, upon which it had previously depended for its raw silk. When the price of fine white silk became prohibitively high because of the export trade, Nanking weavers were forced to look for other, less satisfactory, sources of supply.[89]

Because of its relative proximity to the sources of raw silk, and its receptivity to modern styles and technology, the Soo-chow silk weaving industry survived the difficulties of the late nineteenth and early twentieth centuries far better than Nanking did. Prior to the Taiping Rebellion, Soochow had about 12,000 looms.[90] By 1878 or 1879 it was reported that 5,500 of these had resumed operation, and they produced an average of 83,000 bolts of silk per year.[91] During the late nineteenth century, the weaving of *sha* and *tuan* prospered, and they were sold to markets as distant as Russia, Korea, India, and Burma.[92] There were an estimated 9,000–10,000 looms in operation during the Republican period.[93] Like Nanking, Soochow temporarily suffered from the Boxer Rebellion and the 1911 Revolution.[94] But unlike Nanking, Soochow quickly began to cater to the new Western tastes and styles of the treaty ports after 1911, imitating them as no other sector of the weaving industry could. Soochow's French imitations were sold not only at Shanghai, but as far away as Korea.[95] The craze for Western styles prompted one wag to write:

> Now that the Republic has been newly established, everything is in imitation of Europe and the West. The reverence with which most

young people regard Western products is very profound. The pigtail has been done away with, Western clothes have become fashionable, and there are none among the so-called enlightened girl-students who do not love Western products, to the extent that almost everything they wear from head to foot is a Western product . . . and the powers of both East and West [that is, Japan and the Western nations] have taken advantage of this wonderful opportunity to import every type of product to suit our tastes: all types of Western fabrics, woolens for stockings, hats, shirts, and underwear.[96]

The Hangchow silk weaving industry had experienced a slower recovery from the Taiping Rebellion than either Nanking or Soochow. In 1880, there were only about 3,000 looms, manufacturing about 70,000 bolts per year.[97] During the first decade of the twentieth century, business was slow in the face of competition from foreign fabrics.[98] Around 1909, the city had 4,300 looms, which manufactured about 200,000 bolts of silk.[99] Despite this slow recovery, Hangchow experienced a revival of its fortunes during the Republican period, sometimes called the "golden age" for the weaving industry. Like Soochow, Hangchow was quick to respond to changes in taste, imitating Japanese and French styles.[100]

To meet changing demands, both Soochow and Hangchow quickly adopted modern machinery; the first modern silk weaving factory, the Yi Chang, was established in 1905 at Hangchow. The president of a local industrial school imported iron looms from Japan and hired instructors to teach youths from the northern Chekiang area how to operate them. In 1912 three more factories were established and, in 1914, two more followed. From then on the industry boomed. By 1924, there were sixty factories with 4,000 looms.[101] By the end of the decade there were more than 7,000 looms, producing 250,000 bolts of silk a year and employing a labor force of 13,200 workers.[102] During this boom period, most of the silk at Hangchow was still woven domestically, but it appears that the domestic sector followed the modern sector in adapting to new tastes. About 34 percent of Hangchow's silks were shipped to Shanghai, 15 percent to Southeast Asia, 10.5 percent to Manchuria, and smaller quantities to Hong Kong and other Chinese markets. Less than 4 percent was shipped to the United States or Europe.[103] The revival of the silk weaving industry had a favorable effect on the general economy of Hangchow. It also stimulated the expansion of sericulture in the surrounding areas.[104] An estimated 80 percent of the output of local silk was used by local weavers.[105] Moreover, the weaving boom also encouraged the establishment of steam filatures. In fact, the first steam filatures were attached to the premises of the weaving factories.[106] At Soochow, where there had been 9,000–10,000 looms in operation during the first decade of the twentieth century,[107] seven modern weaving factories, equipped with iron Jacquard looms, had been established by 1918.[108] The first electric power looms were used after the founding of Shanghai's Fu-hwa Factory in 1915.[109] By the 1930s, there were over 2,000 power looms.[110]

The greatest problem encountered by the Soochow and Hangchow weavers in the 1920s and 1930s was competition from foreign synthetic fibers and fabrics, which were extremely popular.[111] At first the guilds resisted the use of synthetic fibers although some Hangchow weavers used them surreptitiously, sometimes in combination with natural silk.[112] Since, however, other centers such as Hu-chou and Sheng-tse were already using synthetics, Hangchow eventually succumbed and permitted their use.[113] The decline, however, of the urban silk weaving industry in Hangchow and Soochow could not be postponed forever. The final blow came from the Japanese invasion of Manchuria in September 1931, and the Shanghai war of January–March 1932, both of which cut Hangchow off from its usual markets. By 1933 it was estimated that 5,000 out of 6,000 of Hangchow's looms were idle, and at least 15,000 weavers were unemployed.[114] At Soochow it was reported that only about 2,500 looms remained in operation during the mid-1930s.[115]

Silk weaving in Kiangnan was not confined to these three

cities. Shanghai, for example, became an important center for modern weaving factories in the 1920s. To a certain extent Shanghai's success was at the expense of the other centers, which were already beginning to fail during the 1926–1931 period. In 1921 there were only 9 weaving factories or workshops at Shanghai; in 1929 there were 68, and by 1931, there were as many as 500 with more than 6,000 looms in operation. According to D. K. Lieu, there were several reasons for this trend. First, as a treaty port, Shanghai provided sanctuary from the political and military disturbances in the interior. Second, Shanghai was closer to cheap sources of electrical power, on which the modern factories depended. Third, Shanghai had easy access to imports of artificial fibers, which were becoming increasingly popular. One particularly well-received type of fabric was called *ti,* an artificial silk made of rayon warp and a cotton woof.[116]

Sheng-tse was another important weaving center, which benefited from proximity to the Hu-chou and Chia-hsing silk districts. In 1880, it was reported that there were 8,000 looms in operation.[117] Even as late as the 1930s, over 4,000 households in Sheng-tse were engaged in weaving, mostly as a primary occupation.[118] Although the local weavers made a great variety of silk fabrics, Sheng-tse was particularly known for its lightweight *ch'ou,* which was extremely popular in the 1920s.[119] In the 1930s, Sheng-tse weavers began to weave a silk blend called *Chung-shan ko,* which also became extremely popular at home and abroad.[120]

In Hu-chou prefecture, accessibility to raw silk sources and adaptability to new tastes also played an important role in the success of the weaving industry in modern times. Each locality had its own specialties. At Kuei-an, the local specialties in the late nineteenth century were *ch'ou, mien-ch'ou, sha,* and *pao-t'ou chüan* (silk for kerchiefs).[121] At Te-ch'ing *chüan* and several types of *ch'ou* were woven.[122] At Shuang-lin, weavers seemed to switch back and forth from one kind of silk cloth to another, depending on current fashions.[123] The prefectural city of Hu-chou was famous for two kinds of silk fabrics—*Hu-ts'ou,* a type of crepe, and *mien-ch'ou,* which was woven from waste silk and had the appearance of cotton. *Hu-ts'ou* began to be popular in the 1860s and 1870s, and was widely marketed in north China, the Yangtze valley, and the Canton area, and later abroad.[124] Since it was woven from *fei-ssu,* or coarse silk, which was not exported, there was no competition between the local weaving industry and the reeling of Tsatlees for export.[125] During the early Republican period, the market for *Hu-ts'ou* was threatened by competition from a Japanese silk import called *yeh-chi ko,* but the Chekiang provincial government established schools where students were trained to weave a new product, *Hua-ssu ko,* designed to compete with the Japanese import.[126] In all, the number of looms operating in Hu-chou increased from about 4,000 looms in 1880 to about 13,000 looms, distributed among sixty factories and about 6,000 households in the Republican period.[127] After 1926, however, the Hu-chou weavers began to lose out to the Shanghai mills, and by 1932, there were only about 4,000 looms operating, producing an annual output of about 300,000 bolts of silk.[128]

Hu-chou's weaving skills spread to the Tan-yang region in Kiangsu, where silk weaving—especially the weaving of *Hu-ts'ou*—was introduced by migrants from Hu-chou in the late nineteenth century. The local product was appropriately called [Tan]*-yang-ch'ou,* and became extremely popular because it could be sold at a lower price than the original *Hu-ts'ou.*[129] At the beginning of the Republican period, Tan-yang also began to manufacture the popular *Hua-ssu ko,* and modern looms began to replace the old wooden looms. By 1924 there were more than 4,000 looms, producing 300,000 bolts a year.[130]

Not all weaving centers in Kiangnan, however, were so successful. Several older centers experienced irreversible decline after the Taiping Rebellion—for example, P'u-yuan *chen* in T'ung-hsiang hsien,[131] Chia-hsing,[132] Shao-hsing,[133] and Chin-kiang.[134]

One Japanese study estimated that, during the mid-1920s, the total amount of raw silk woven in China was about 120,000 piculs, and possibly much higher. Of this amount, half, or 60,000 piculs, was used in the Kiangnan region—20,000 at Hangchow; 20,000 at Sheng-tse and Hu-chou together; and another 20,000 at Soochow, Nanking, and Shanghai.[135] The foregoing survey of the major silk weaving centers in Kiangnan confirms the notion that the silk weaving industry, as a whole, after suffering a mid-nineteenth century net decline, managed to maintain a steady domestic market until the 1930s. However, the experiences of individual centers varied widely.

Although Shen Lien-fong, head of Shanghai's cocoon guild, complained that the expansion of the modern silk reeling industry had been blocked by strongly entrenched weaving interests,[136] it appears that the reality was not so simple. At Nanking, for example, the reverse was true; the relative inaccessibility and costliness of raw silk contributed greatly to the decline of its weaving industry. On the other hand, it may well be argued that the raw silk export trade might have expanded more easily if the weaving industry had not been relatively strong in other Kiangnan centers. It seems clear that one of the crucial factors which determined the ability of a weaving center to survive was not simply its proximity to sources of raw silk, but also the extent to which it could shift its technology and artisanship to meet new tastes. The structure of the late nineteenth and early twentieth century domestic demand for silk fabrics was quite different from that of the early Ch'ing period. No longer was there such a large market for the ultrahigh quality luxury brocades and satins; a larger part of the market was geared toward less expensive and less luxurious grades of silk. The popularity of synthetics and lower grades of silk suggests the emergence of a new kind of market, probably based on the tastes of a more Westernized treaty port population. This suggests a reason why the smaller urban weaving centers, and the rural areas, may have suffered less dislocation in their weaving activities than the three major urban industries. Since the smaller centers, such as Sheng-tse and Hu-chou, wove much of their fabrics from lower grades of silk, they were not in direct competition with the raw silk export trade, which dealt primarily in high quality Tsatlees. This would also explain why the major weaving centers outside of Kiangnan survived as well as they did; the Pongees of Shantung used wild silk, and the fabrics of Szechwan used yellow silk.

That the weaving industries in the three major cities of Nanking, Soochow, and Hangchow did experience severe dislocation is corroborated by the activities of the silk guilds after the Taiping Rebellion, and the high incidence of strikes in the twentieth century. Silk guilds had a very old history in both Soochow and Hangchow,[137] but after the Taiping Rebellion, their numbers proliferated and the previous distinction between residential associations (*hui-kuan*) and professional guilds (*kung-so*) began to overlap.[138] As P'eng Tse-i has observed, the proliferation of guilds reflected greater specialization of function, as well as the heightened tensions created among different interest groups by foreign and domestic competition, technological change, and external dislocation.[139] Table 22 shows the fragmentation of Soochow silk guilds. At Hangchow, the major silk weaving guild hall, colloquially known as the Kuan-ch'eng t'ang, was rebuilt by local leaders after the Taiping Rebellion, but after the turn of the century, the pressures to modernize resulted in the creation of several splinter groups. While the *chang-fang* and the traditional interests remained in the old guild, some weavers and representatives of the modern factories formed separate guilds.[140]

Both Soochow and Hangchow were disturbed by an increasing number of weavers' strikes starting in the 1890s and continuing through the 1910s. Most of the strikes were an expression of the weavers' anger at being laid off work by their *chang-fang* employers, but there were also some strikes over wages. As Kojima Yoshio has pointed out in his study of these strikes, they reflected not only general economic difficulties, such as inflation and also the 1910 financial crisis at Shanghai,[141] but also a

TABLE 22 Silk Guilds at Soochow

Name	Function/Composition	Date of Origin
San-shan hui-kuan[a]	raw silk, *ch'ou*, bamboo	Ming Wan-li period
Wu-chün chi-yeh kung-so[b]	weaving	1740 (orig. 1295)
Ssu-yeh kung-so[a]	raw silk	unknown, changed name to kung-so in 1870
Ching-pang ching-yeh kung-so[a] (orig. Yuan-ning hui-kuan)	warp	unclear, first mentioned in 1879
Wen-hsuan kung-so[a]	dyeing of silk	unclear
Wen-chin kung-so[a, b]	for independent weavers of *sha* and *tuan*	unclear, after 1911
Ssu-pien kung-so[a]	silk braid	unclear
Jung-yeh kung-so[a]	velvet	unclear
Ching-yeh kung-so[a]	warp (Chen-tse clique)	unclear
Ch'i-hsiang kung-so[b]	*sha, tuan, ch'ou, ling* (both *chi-hu* & *chi-chiang*)	1839
Yun-chin kung-so[b]	*chang-fang* and *chi-hu*	1860–1870
Hsia-chang kung-so[b]	for workers only, *kung-chiang*	1910

Sources: [a]Liu Yung-ch'eng, "Shih-lun Ch'ing-tai Su-chou shou-kung-yeh hang-hui," *Li-shih yen-chiu* 1959. 11:23–24 (November 1959).
[b]P'eng Tse-i, "Ya-p'ien chan-cheng ch'ien Ch'ing-tai Su-chou ssu-chih-yeh sheng-ch'an kuan-hsi ti hsing-shih yü hsing-chih," *Ching-chi yen-chiu* 10: 69, 72 (October 1963). See also *Shina keizai zensho*, comp. Toā dōbunkai (Tokyo, 1908), XII, 279, which lists Chi-chuan kung-so for tax collecting.

growing class or group consciousness on the part of weavers against their employers, as well as their increasing dependence on a broader market. The manner in which these strikes were usually settled, however, shows the persistence of the traditional symbiotic relationship between the guilds and local authorities. If anything, the political functions of guilds had expanded since the Taiping Rebellion, as they assumed responsibility for the collection of the likin (*li-chin*) tax on the internal transit of commodities, and other duties, and more responsibility for the settlement of disputes.[142]

In short, the modern fate of the silk weaving industry in Kiangnan was very mixed. The disturbances among weavers in Kiangnan cannot be attributed simply to the competition of the silk export trade. Changing markets, sources of supply, technology, and political climate all played a role. Yet in the final analysis, one cannot help being impressed by the strength and durability of the silk weaving industry and the domestic demand that it represented. While this demand did not prevent the growth of sericulture and silk reeling in new areas, in the traditional weaving centers of Kiangnan, it seems to have exercised an inhibiting influence on the maximal development of the silk export trade.

FIVE

Foreign Trade and the Rural Economy

The development of the silk export trade had a measurable impact on the rural economies of the regions that produced silk, and yet despite the frequent and self-conscious exhortations of nineteenth-century bureaucrats to develop sericulture, there were limits to growth, particularly in the old areas. In the twentieth century the modern filature industry's most frequent complaint was the inadequate supply of high-quality cocoons; there seemed to be obstacles to both quantitative increase and qualitative improvement which could not be surmounted. It was not just the weaving industry which prevented a stronger response to the needs of the export trade, but also the agricultural and commercial practices of the traditional silk regions themselves.

BUREAUCRATIC MYTHS AND THE
PROMOTION OF SERICULTURE

There is no question that the silk export trade had extremely beneficial effects on the economy of the Hu-chou and Chia-hsing areas, particularly just after the Taiping occupation. Robert Fortune in visiting Hu-chou in the late 1850s had observed, "I have visited many Chinese towns, and I must say I never saw the people as a whole better dressed than those of Hoo-chow. Every person I met above the common working coolie was dressed in silks or crape, and even the coolies have at least one silk dress for holiday wear."[1] Foreign trade made it possible to maintain this level of prosperity after the Taiping Rebellion. The merchants of Nan-hsun in particular benefited from the trade. It was said that the slightly rich became medium rich, while the medium rich became very rich. There was another saying that all of the wealth of the prefectural city of Hu-chou could not match half the wealth of the town of Nan-hsun. Many of these rich Nan-hsun families invested their profits from the silk trade into salt, pawnshops, and wastepaper enterprises, as well as in land.[2]

Since ancient times sericulture had been regarded as the foundation of the peasants' livelihood and, hence, the fiscal well-being of the state. Official ideology and popular custom both held that it was the man's duty to farm the land and the woman's duty to weave.[3] It was the official's duty to encourage the spread of sericulture and silk weaving among the people. Traditional Chinese historiography attributes the flourishing of sericulture in the Ming dynasty to the wise policies of its founding emperor, T'ai-tsu, who decreed that all families with five to ten mou of paddy fields should plant a half mou each of hemp, cotton, and mulberry. Those with more than ten mou of land were to double their quotas. Those who did not plant these crops would still have to weave one bolt of silk cloth to satisfy their tax obligations. In 1394 the people were ordered to plant even more mulberry. In every hamlet, two mou of mulberry seedlings were to be planted and later divided up among the local households for transplanting.[4] Although such pronouncements were honored more in their breach than their acceptance, they nevertheless set a tone which provincial and local officials tried to emulate.

During the late Ming period, a number of gentry scholars in the Kiangnan area wrote agricultural and sericultural manuals in which they praised the advantages of sericulture. One of the most famous of these manuals was the *Pu Nung-shu,* which was an amplification of a work called the *Shen-shih Nung-shu,* by a Mr. Shen, whose exact identity has never been established. Chang Lü-hsiang, the author of the *Pu Nung-shu,* and a native of T'ung-hsiang hsien in Chia-hsing prefecture, emphasized the increasing economic importance of sericulture, saying that mulberry cultivation was in fact more important than rice cultivation for his locality.[5] Mulberry cultivation, in Chang's view, had several important advantages. It required less human labor and less intensive cultivation of the land than rice. It could be done at a more leisurely pace. And it was less severely affected by drought than rice was.[6] In a rather cryptic, but well-known, passage, Chang Lü-hsiang wrote:

> In T'ung-hsiang, paddy fields (*t'ien*) and dry fields (*ti*) are equivalent, and sericulture is very profitable. Chia-shan, P'ing-hu, and Hai-yen to the east, and Kuei-an and Wu-ch'eng on the west, all have more paddy fields than dry fields. [However,] in any *hsiang* agricultural activity reveals that the profits from dry fields are more ample, and planting more rice fields is not as worthwhile as cultivating the dry fields. . . .There is a popular expression, "It takes a thousand days to start a rice paddy, but only one day to start a mulberry field." Moreover, although in the most fertile years, the paddy fields will yield three piculs of rice and one and a half piculs of spring wheat per mou, these years are rare. Most of the time three piculs is the average. In the most prosperous years, the dry fields will yield enough leaves per mou to feed ten baskets of silkworms, or in lesser years, four or five, or at the very least, two or three. When rice is cheap and silk is expensive, one basket of silkworms can match the profit from one mou [of grain]. . . .There is a popular expression,

"If you plant mulberry for three years, you can pick leaves for a lifetime." Never was there a case of not getting a lifetime of ease for one effort.[7]

By saying that in T'ung-hsiang paddy fields and dry fields were equivalent, Chang did not mean that they were equivalent in acreage, but that the profits derived from rice and mulberry were equivalent since the value of the yield of the dry fields was much higher than that of rice. This passage presents several problems. In agreeing that "it takes a thousand days to start a rice paddy, but only one day to start a mulberry field," Chang was disregarding the fact that mulberry trees took about seven years to reach a productive maturity. Furthermore, Chang was not taking into account the relative amounts of labor necessary to cultivate rice and mulberry fields. The labor cost of raising silk is in some sense the rice which is not raised as a consequence.[8]

Yet Chang's observations were consistent with the growing recognition of the importance of cash crops and the trend in the Kiangnan area during the Ming toward the commercialization of agriculture. The 1617 (Wan-li) edition of the T'ung-hsiang gazetteer stated that the profits from beans and mulberry were four, and sometimes even five, times that from rice cultivation.[9] As summarized in Table 23, calculations made by Ch'en Heng-li in his study of the *Pu Nung-shu* show that, in the late Ming and early Ch'ing, dry fields in T'ung-hsiang represented 16 or 17 percent of the total cultivated acreage. In nearby Ch'ung-te hsien (renamed Shih-men after 1912), dry fields increased from 12.46 percent in the Wan-li period to about 41.42 percent in the K'ang-hsi period. Ku Yen-wu, the eminent seventeenth-century scholar, wrote that in Ch'ung-te rice paddies and dry fields were equal in importance. The rice harvest, he said, was sufficient to feed the people for only eight months of the year; for the remaining months, the people could depend only on sericulture to supplement their incomes.[10]

During the early and mid-Ch'ing, official efforts to encourage the spread of sericulture continued apace. There were attempts

TABLE 23 Mulberry Acreage as Percentage of Total Cultivated
Land in Chia-hsing and Hu-chou

Hsien	1581^a	1667–1713	$1930s^b$	1950s
T'ung-hsiang	16.27	16.99	32	52.42
Shih-men (Ch'ung-te)	12.46	41.42	26	44.61
Kuei-an, Wu-ch'eng (Wu-hsing)	11.87	——	36	——
Chia-hsing	7.16	——	21	24.68
P'ing-hu	14.84	——	22	7.17
Chia-shan	1.81	——	2	3.94
Hai-yen	——	——	25	——

Sources: [a]With the exception of the 1930s column, these estimates are from Ch'en Heng-li, pp. 120–123, 253. Strictly speaking, Ch'en's calculations represent paddy fields as opposed to dry fields, and he simply assumes that the dry fields were comprised entirely of mulberry. Since common sense tells us that this could not have been the case, all of Ch'en's figures for dry fields should be considered higher than the actual figures for mulberry fields would be.

[b]*Chung-kuo shih-yeh chih: Che-chiang sheng*, IV, 167–168. The same material is found in Yüeh Ssu-ping, pp. 18–19. Unlike Ch'en Heng-li's calculations, these estimates refer specifically and only to mulberry fields and therefore are lower than an estimate for total acreage of dry fields would be. For example, Ch'en calculates that dry fields in Wu-hsing in the 1930s were 40.44 percent of total cultivated acreage, rather than 36 percent.

to introduce sericulture to new localities in various provinces.[11] But after the Taiping Rebellion, the pace of such activity accelerated as the new opportunities for foreign trade developed and, within the Kiangnan area, a pattern was repeated. Local representatives were deputed by the magistrate to go to Hu-chou to buy *Hu-sang* saplings to distribute to the peasants. Very often a mulberry or sericulture bureau would be set up to provide instruction to the peasants on both mulberry cultivation and sericulture with experts from Chekiang as resident instructors. Considerable attention was devoted by officials and scholars to the writing and revision of sericultural manuals in order to disseminate knowledge of sericultural techniques.

For example, after the Taiping Rebellion, the magistrate of Tan-t'u, Shen Ping-ch'eng, established a mulberry bureau for the purpose of instructing the people how to plant *Hu-sang*. According to the local gazetteer, his efforts were well rewarded, and mulberry fields became widespread within the hsien.[12] Shen was a native of Kuei-an in Hu-chou, a holder of the *chin-shih* degree, and a member of the statecraft (*ching-shih*) school. He later served as governor of Anhwei province from 1888 to 1894. During the T'ung-chih period he authored, or at least sponsored, the publication of a sericultural manual entitled *Ts'an-sang chi-yao*, which was probably the basis of *Ts'an-sang shuo*, which in turn formed the basis of *Kuang ts'an-sang shuo chi-pu*, compiled by Chung Hsueh-lu, and sponsored by Tsung Yuan-han.[13] The latter was prefect of Yen-chou prefecture in Chekiang, a native of Hu-chou, and an editor of the 1874 edition of the Hu-chou gazetteer. He related in the preface of his work how he tried to introduce sericulture to Yen-chou by ordering some of his servants to grow silkworms in the court-yard of his yamen, and recruiting an expert from Shao-hsing to set up a sericultural bureau in the following year. He also had several thousand mulberry saplings imported from Hangchow and had them planted. In Yen-chou, he said, the land was hilly, and the wet fields were scarce. Whenever the weather was bad, the peasants suffered. With sericulture, the women could bring in an additional income from weaving.[14]

Another example was that of the magistrate of Ching-chiang hsien in Kiangsu, named Huang Shih-pen, a native of Ch'ien-t'ang in Hangchow prefecture, who also purchased mulberry saplings from Hu-chou and tried to teach the local people seri-cultural techniques from Chekiang. In connection with his pro-ject, he wrote a sericultural manual called *Ts'an-sang chien-ming chi-shuo*, which was first published in 1882.[15] Chang Hsing-fu, the author of another sericultural manual, *Ts'an-shih yao-lueh*, was a native of An-chi hsien in Hu-chou fu, and received the *chü-jen* degree in 1870. He was co-editor of the 1874 edition of the An-chi gazetteer, as well as the author of several scholarly treatises on the ancient dictionary *Shuo-wen*.[16] Chang empha-sized that the most important step in introducing sericulture to a new area was to invite experts to instruct the local people. If possible, several families from the silk districts should be invited to set up sericultural stations in the outlying areas, in places accessible to the peasants. As soon as one locality got good results, the others would want to follow its example. He emphasized that, since sericulture was an enterprise of the ordinary people, stations should not be set up in towns where their usefulness would be limited, but on the outskirts. He also advised that these experiments be supervised or sponsored by the local gentry (*hsiang-tung* or *shen-tung*) rather than the local officials, who were feared by the people.[17]

The promotion of sericulture was not restricted to relatively obscure local officials. There was hardly a high Ch'ing official in the late nineteenth century who did not at some point at least pay lipservice to sericulture. Both Tseng Kuo-fan and Shen Pao-chen, when they served as officials in Nanking, tried to promote the revival of the silk trade there after the Taiping Rebellion.[18] And wherever Tso Tsung-t'ang served as governor or governor-general, he tried to introduce sericulture.[19] Li Hungchang and Chang Chih-tung were also promoters of sericulture in the latter part of the century.[20] Chang Chien, important leader of the provincial assembly in Kiangsu and promoter of local projects, wrote that when he saw how Wusih had profited from the cocoon business, he arranged to have the likin on silk repealed for ten years near his local area of Nan-t'ung and Haimen and to have a few cocoon hongs established.[21]

The record of failures was, however, almost as impressive as the record of initiatives. Tso Tsung-t'ang's efforts in Fukien, for example, met with failure because the climate and soil were unsuited for mulberry cultivation. In Sungkiang and T'ai-ts'ang, and other Kiangnan locations near the ocean, sericulture also failed because the soil and climate were unsuitable. In Sungkiang, despite the fact that cotton was the main crop, some gentry and officials waged persistent campaigns to popularize sericulture at the turn of the century and in the early part of the twentieth.[22] Although mulberry fields were planted as model projects, sericulture never caught on, and the output of silk from the area was minimal. Similar failure occurred in T'ai-ts'ang despite good intentions.[23]

Failures were probably more frequent in north China. One keen observer was T'ao Pao-lien, a native of Hsiu-shui in Chia-hsing. While traveling with his father, who was en route to assume his new post as governor of Sinkiang in the 1890s, he wrote a record of his impressions of north China. T'ao criticized the attempts to introduce sericulture indiscriminately, particularly to the north where conditions, he felt, were not suited to sericulture but were admirably suited to cotton cultivation. His essay was very revealing of both the official myths and the scientific and economic realities concerning sericulture. Briefly, his arguments were that the dry and dusty climate of the north was not a hospitable environment for mulberry trees or silk-worms. Moreover, the people of the north could not adapt to the extremely labor-intensive and seasonal aspects of sericulture. In his native place, Chia-hsing, commerce was relatively advanced and therefore division of labor and specialization were possible. Each person need learn only one skill—reeling, weaving, woof- and warp-making—and own only the tools necessary for that skill. But in north China, each family would have to learn all the steps in sericulture. In Chia-hsing, where nearly all the households raised silkworms, the losses and gains could be spread around, but in north China a loss in one season would be difficult to recover in the next. Finally, in south China, where both rich and poor alike wore either silk or cotton garments, there was always a market for silk products, but in north China, where everyone was more frugal, the market was not so reliable.[24]

The promotion of sericulture by Ch'ing officials was an expression of the vitality of local leadership, and yet, at the same time, their efforts were largely unsuccessful, because they were strictly local and ad hoc in nature and not coordinated into a national plan or policy. Not only was the magistrate's jurisdiction local, but his perspective was too. His motive for encouraging the development of sericulture, or the introduction of any new crop, was to increase the economic wealth of the locality. As Yen Chung-p'ing has observed, at the foundation of the traditional concept of economic well-being was the notion of the self-sufficiency of each household. Weaving and sericulture were encouraged because they would help to make the household economically independent.[25] Although the expansion of the silk export trade should ultimately have benefited the nation as a whole, neither the concept of a national economic good, nor that of using foreign trade to the nation's advantage had gained much acceptance by the end of the Ch'ing. At the very time that these local efforts were being made, the central government took no action to promote the silk industry. Robert Hart, the Inspector General of the Maritime Customs, lamented the refusal of the Tsungli Yamen in 1890 to enforce the Pasteur system of inspection, "They're queer fellows these Chinese?—I verily believe they are saying—'What a good thing it would be to follow advice and rehabilitate tea and silk—but might it not be a better thing to leave them alone and perhaps thereby get rid of the foreigners who came here for them?'"[26]

ECONOMIC REALITIES AND THE PEASANT HOUSEHOLD

Although official rhetoric claimed that sericulture could be done by every household, and mulberry could be planted in any soil, the economic reality was that, except in Kwangtung, no household could afford to become overly specialized in sericulture, and no locality could afford to devote too much land to mulberry planting. In most areas it was rare for a peasant household to devote all of its land to mulberry, and it was rare to have fields that were planted exclusively in that one crop. The typical pattern in Kiangnan, as Robert Fortune observed, was to plant mulberry along the edge of fields on high ground, while rice was planted in low fields. Even in the Kiangnan silk region, which extended from Hai-ning and Hangchow in the south, to Hu-chou in the north, and across to the other shore of the T'ai-hu, mulberry fields constituted only about 30–40 percent of the total cultivated acreage according to one estimate.[27] As Tables 23 and 24 show, even localities such as Wu-hsing devoted only 36 percent of their land to mulberry. In Wusih, where mulberry was more often grown in plantations rather than in scattered fashion, mulberry fields in the 1920s and 1930s represented only about 20–30 percent of all cultivated land in the hsien.[28] Typically, each household had no more than one or two mou of mulberry land. Rice and wheat were still the major crops in the Wusih area.[29]

Why was this the case when Chang Lü-hsiang's claim that mulberry cultivation could be three times more profitable than rice cultivation was in fact reflected in the higher rents charged for mulberry fields as opposed to rice fields? In Shao-hsing in the 1920s, for example, one mou of rice paddy land would rent for 70 to 100 yuan, while one mou of mulberry land would cost 100 to 150 yuan. At Wusih a mou of rice paddy would rent for 50 to 80 yuan, while a mou of mulberry land would cost between 70 and 150 yuan, averaging 80 yuan.[30] In Ch'ang-shu hsien in Soochow prefecture, one mou of medium-grade rice paddy would cost 70 to 80 yuan, while a mou of average mulberry land would cost 120 yuan and a mou of high-grade mulberry land would cost 200. By contrast, a mou of ordinary dry field, suitable for tobacco for example, would cost only 50 yuan.[31]

For a locality, the major constraint on overspecialization was its need for rice. In Hu-chou and other silk areas, the shortage of rice for local consumption was a constant threat to local welfare. The Nan-hsun gazetteer decried the fact that sericulture had expanded at the cost of rice cultivation, a particularly serious problem when the rice harvest was poor.[32] The Te-ch'ing gazetteer said that the profits from sericulture could easily be wiped out if the rice price got too high; the bitterness of sericulture, it noted, was at least as great as its potential

TABLE 24 Mulberry Acreage and Households in Sericulture
in Chekiang c. 1933

Place	Mulberry Land (in mou)	Percentage of Total Cultivated Land	Percentage Households in Sericulture
HU-CHOU FU			
Kuei-an			
}Wu-hsing	545,569	36	87
Wu-ch'eng			
Ch'ang-hsing	97,000	12	65
An-chi	7,500	8	43
Hsiao-feng	7,500	4	41
Wu-k'ang	18,309	15	45
Te-ch'ing	76,654	37	86
CHIA-HSING FU			
Chia-hsing	350,450	21	74
T'ung-hsiang	182,945	32	90
Shih-men	53,000	26	85
(Ch'ung-te)			
Hai-yen	97,700	25	81
Hsiu-shui	––	—	—
P'ing-hu	76,080	22	64
Chia-shan	9,000	2	16
HANGCHOW-FU			
Ch'ien-t'ang			
}Hangchow	355,596	18	74
Jen-ho			
Yü-hang	58,200	21	83
Lin-an	36,600	12	64
Yü-ch'ien	9,900	9	20
Ch'ang-hua	3,800	2	15
Hsin-ch'eng	13,250	11	46
Fu-yang	18,000	7	15
Hai-ning chou	350,000	35	89

TABLE 24 (continued)

Place	Mulberry Land (in mou)	Percentage of Total Cultivated Land	Percentage Households in Sericulture
SHAO-HSING FU			
Hsiao-shan	57,280	7	62
Chu-chi	94,000	8	85
Hsin-ch'ang	28,500	5	32
Yü-yao	3,400	—	—
Shang-yü	8,400	0.5	12
K'uai-chi			
}Shao-hsing	9,400	0.5	3
Shan-yin			
Sheng	74,660	6	75
Other	—	—	—

Source: Chung-kuo shih-yeh chih: Che-chiang sheng, IV, 167–168, 184–185. Compare these estimates with those of Yüeh Ssu-ping, pp. 169–170.

benefits.[33] If a locality could not import rice cheaply, it could not afford to specialize in mulberry and sericulture.[34] The Kiangnan region was blessed with an intricate system of waterways, which provided a cheap form of transportation. Rhoads Murphey has observed that "the leading silk areas within the delta coincide closely with the areas where navigable waterways are most numerous."[35] Just as the completion of the Shanghai-Nanking railway in 1908 was an important factor in the eventual rise of Wusih as a silk filature center,[36] the completion of the Shanghai-Hangchow-Ningpo railway around 1916 was a factor in encouraging the spread of sericulture to new parts of northern Chekiang.[37]

Competition from other cash crops probably served as less of a direct limitation on mulberry cultivation than the need for rice. Although cotton and mulberry were sometimes grown in the same field, basically cotton required alkaline soil.[38] In Kiangnan the main cotton-growing centers—T'ai-tsang and

Sungkiang prefectures—were distinct from the main sericultural centers. Tea required hilly land, and hemp was much less profitable than mulberry.[39] On the other hand, wheat was sometimes planted in the same field as mulberry. And tobacco, which became an important cash crop in the twentieth century, may in some localities have limited the growth of sericulture.[40]

For an individual peasant household, the constraints on overspecialization were quite specifically related to the characteristics of mulberry cultivation and sericulture as economic activities. The first characteristic was that sericulture involved a high degree of natural risk, which in turn generated a high degree of commercial risk. Mulberry trees required a great investment of time as well as land. The sericultural manuals pointed out that if one got discouraged before the six or seven years necessary for a tree to mature had elapsed, all the previous effort could be wasted.[41] "Mulberries are difficult to grow to maturity and easily fail. The first year you plant them. Within three years, the fruit will come out. Only six years later do you get a flourishing tree. If you don't look after it diligently, the pests will not be driven away. If you don't cultivate the soil weekly or semi-monthly, the color will be dull."[42]

Although the unpredictability of the weather affected all crops to some degree, the price of mulberry was further affected by speculation about domestic or foreign market conditions. The price of leaves could fluctuate so erratically that there was a popular expression in Kiangnan that "even the gods find it difficult to predict the price of leaves."[43] Often the peasants would resort to divination to predict the price.[44] Although it is clear from such sources as the *Pu Nung-shu* that the price of mulberry leaves in the seventeenth century was exceptionally unstable compared to the price of rice,[45] the fluctuations in the nineteenth and twentieth centuries were even wider. Many contemporary sources attest to the fact that prices could shift by a factor of ten or more. One source said that at Hu-chou the price of leaves could rise as high as 3,000–4,000 *ch'ien* (cash) per picul, or drop as low as 200–300 *ch'ien*.[46] The price of leaves fluctuated within a given season, as well as from one season to the next. It could vary even from dawn to dusk of the same day. One passage describes the trials of a silk-producing family in Nan-hsun:

> When our family sold leaves in the winter month, one *tan* [picul] was worth just 500 *ch'ien*. In the silkworm month [the fifth lunar month], the price was up to one thousand. But by the time of the third repose, there was a sudden reversal, and the price dropped to one *ch'ien* per *chin* [or one hundred *ch'ien* per picul or *tan*]. When your husband heard this, he scolded you, but it is useless to have regrets. . . . What our family lost cannot be calculated. . . . Coming home all I can do is to cry.[47]

While these figures may be more literary than real, they at least suggest the problem and magnitude of the fluctuation. After the Taiping Rebellion, the price of leaves soared steeply to about 5,000 *ch'ien,* or five taels, per picul.[48] Similar conditions prevailed in the twentieth century. At Wusih, in a good year, the price would rise from two or three yuan per picul to eight or ten yuan, but in a bad year it could fall to twenty cents per picul. According to D. K. Lieu, the price of leaves reached its height around 1921, reaching 4.50 yuan a picul at best, 1.50 yuan at worst. In the 1930s the price dropped, ranging from three yuan at the very best to thirty to forty cents in the worst year of 1934.[49]

There is no doubt, however, that in a good year, mulberry leaves could be a very profitable enterprise. In Hu-chou it was said that the cost of cultivating and fertilizing one mou of mulberry trees would not exceed two taels, and the profit would be double that.[50] Of course, if one raised silkworms in addition to trees, the profit would be still greater. In Wu-ch'ing it was said that the cost of maintaining one mou for one year would not exceed three taels, but the usual profit would be double that.[51] In Soochow in the 1920s, the Silk Association found that it cost the individual peasant ten yuan to cultivate one mou of mulberry trees: seven yuan for renting the land and three yuan for fertilizer. On the open market he could get three yuan per picul for his leaves. Since the yield in this area was 800–1,000 catties per mou, he would be able to get twenty-four to thirty yuan per mou, thus earning a net profit or fourteen to twenty yuan.[52] Japanese observers felt the Chinese prices for mulberry were extremely high because of scarcity of mulberry fields.[53]

The price of leaves varied from one area to the next within Kiangnan.[54] Since leaves were easily perishable, there was a limit to the distance which they could be transported. Boats might come from 100 *li* away, but they must have made the trip within one day and one night.[55] In most cases, however, leaves were purchased from nearby markets. The *Hu ts'an-shu* said that there were mulberry markets within 30 *li* to the southeast of Wu-hsing and Wu-ch'eng. Nan-hsun itself produced only enough leaves for the spring crop and had to buy additional leaves from Chia-hsing. Shih-men, T'ung-hsiang, and T'ung-t'ing were said to produce the most leaves. Most of the mulberry leaf brokers were concentrated at Wu-chen.[56]

There were several types of commerce in mulberry leaves, all of which were of early origin. One involved the renting or reserving of trees in advance, usually before the lunar new year. This was called *miao-yeh,* or *miao-sang.* In Hu-chou people who did not have their own mulberry trees generally rented them from others. They could either pay in advance, or pay the amount plus a surcharge after the leaves were picked.[57] Because of the uncertainties of the leaf market, renting trees was considered less risky than buying leaves on the open market.[58] As it was practiced in Soochow in the 1920s, the buyer was entitled to pick the tender leaves for feeding the worms in the early stages.[59]

To purchase the leaves directly was called *shao-yeh* (lit., "leaf sale"). There were two methods of *shao-yeh.* In the *hsien-shao* (lit., "cash sale") system the leaves were sold on the open market for cash when they were fully mature. The leaf market opened after the final repose and was called *san-shih* (lit., "three markets"), because it consisted of three stages: the first lasted three days, the second, five days, and the third, seven days.[60] The sales of leaves were conducted by the leaf hongs, often called *ch'ing-sang yeh-hang*. They acted as middlemen between buyer and seller and usually exacted a commission from each, but sometimes just from the seller.[61] *Hsien-shao* involved a big risk, since the price could fluctuate wildly during the ten days between the last molting and the spinning of the cocoon.[62] Usually the peasants would first buy half the quantity they thought they would need, guarding against the possibility that the worm crop would not be good that season.[63] The leaf market was apparently the scene of many small dreams. The rich got richer, while the poor peasants got poorer, according to the popular literature. "There are those who go broke because the price of the leaves is too low; there are those who do not sell fast enough before the peak time is over; and there are those who get cheated by the brokers from the city and do not know where to recover their capital."[64]

The second method of *shao-yeh*, called *she-shao* (lit., "credit sale"), involved making an oral contract to set the price of the leaves in advance. Payment would be made at the end of the season after the silk was sold.[65] This method was considered advantageous to the buyer, because he would certainly get a fair price in advance. At the time of completing the contract, he also agreed on the time for picking, usually three to five days after *li-hsia* (May 5). If this agreement was not made, the buyer would want the leaves earlier, while the seller would want to wait until later, when the leaves would weigh more.[66] By the early twentieth century, leaf-hong merchants would take their orders at the end of the winter of the previous year. At that time the price of the leaves would be the lowest, since the peasants were trying to settle their debts for the year.[67] This was still done in Soochow in the 1920s, and it usually meant that the peasant was selling below par because of financial need. Usually half the agreed amount was paid at the time of concluding the contract, and half was paid at the time of delivery.[68] The advance-contract system was more common in the villages, while the open-market system was used more in cities.[69]

Although safer than selling directly on the open market, *she-shao* had its own risks. In 1739 Hu Ch'eng-mou wrote:

> Among the people of Wu-chen there are those who don't raise silk-worms, but do business. This is called "scheming." Although it resembles commerce, actually it is gambling. In reality those who lose money are many, and those who gain a profit are few. There are those who go broke and have no way to make restitution. There are those who get a profit, but upon returning to Wu-chen find their fortunes turned to complications. When the [price of] leaves is high, then the seller defaults [because he can get a better deal elsewhere], and the buyer not only has no leaves to distribute but also has no way to get back his capital. If he has already sold out [his leaves] and in turn needs to buy some more as a supplement, and if the leaves are cheap, the buyer defaults [because he can get them cheaper than his contract price elsewhere], and the seller not only has no place to get his money, but he has to prepare to lower the price of the leaves and seek to sell them [at a loss]. In this case, the creditor will pursue the family members and pick a fight with them, which sometimes develops into a capital case. Therefore, the matter of selling fresh [leaves] is an evil among our local customs.[70]

When the price of leaves was very high, it would be good for the seller of leaves but might be disastrous for the raiser of the silkworms, who might have to destroy some of his silkworms.[71] In the old days this was called having "empty-headed silk-worms."[72] There were many stories of the hardships caused by the market's unpredictable changes. In north Kiangsu, around 1890 there was a young girl whose family had raised an especially large quantity of silkworms. Since they had no money to buy leaves, they had to sell their oxen. But later they discovered that they were still short of leaves. The family wanted to discard the worms, but the daughter insisted on exploring every last possibility. While she was outside looking for mulberry leaves, her father buried all the silkworms. When the daughter learned of this, she felt there was nothing left to live for, and so she hanged herself.[73] In Mao Tun's "Spring Silkworms," old Tung Pao and his family decided, after heated discussion, to pay an exhorbitant price for extra leaves when it appeared that they had an exceptionally fine crop of silkworms. But when the local steam filatures closed down, they were unable to sell most of their cocoons—thus suffering the loss of the capital they had invested.[74]

The high degree of risk involved in sericulture made it impractical for a household to devote too large a portion of its land to mulberry or to attempt to raise too many silkworms. As the Ch'ing manuals said, if the family was too greedy and had too many trees, it might be ruined if the price fell too low. If the family had too many worms, it would have to guard against leaves' becoming too expensive. If the family had too many of both trees and worms, it might not have sufficient labor.[75] The ideal situation was to cultivate just enough trees to have a certain supply of one's own, and depend on the market for any extra one might need. Since the largest portion of the expense of sericulture was represented by the cost of leaves—at least 50 to 60 percent, or in some cases as high as 94 percent—the element of risk was transferred to each subsequent stage of silk manufacturing.[76]

Beside the element of risk, the second major characteristic of sericulture—outside of Kwangtung—was its seasonal nature. One reason why sericulture and mulberry cultivation—when done on a small scale—complemented rather than conflicted with the requirements of other crops was that the sericultural season meshed very well with the seasonal demands of rice and other crops. Mulberry harvesting and the sericulture month took place in the early spring and were completed before rice had to be sown. Having only one major silkworm generation a year then freed labor to plant rice and harvest it in the fall and thus took advantage of seasonally underemployed labor.[77] If one locality specialized too heavily in mulberry, or any other crop for that matter, labor demands could not have been so evenly spread out over the year.

A third characteristic of both mulberry cultivation and sericulture was that they were extremely labor-intensive activities.

J. Lossing Buck's study of eighteen localities in the Yangtze rice-wheat area found that the man-labor requirements for mulberry were higher than for any other crop—76 days per crop acre were required for rice, 126 days for tea, but 196 days were required for mulberry.[78] If the scale of mulberry cultivation by each individual household was restricted to that which could be handled by the family members themselves, they could save the cost of hiring extra labor—which in the case of mulberry could constitute more than one-third the total cost of working one mou of land.[79] The same principle applied to the scale of sericulture within each household. The demands of the silkworm month were so intense that family members would lose many nights' sleep while they tended the growing worms. If the family raised more worms than it could take care of by itself, the cost of hiring outside help might wipe out any profits that might be made from selling cocoons or silk. Next to the cost of leaves, the cost of labor was the second largest cost of raising silkworms, ranging from 30–50 percent of total cost in some cases.[80] Since it was the custom in the Kiangnan area to reel silk from live cocoons within a short period of time, the number of cocoons that a household could reel without hiring help was limited by the number of available hands at home. In reeling silk, hired help was used more frequently than in sericulture, in part because of the time pressure, the need for some degree of skill, and also because domestic reeling machines often required male workers. But because the cost of hired help was very high,[81] as a rule, each household would raise only as many silkworms as its members could handle by themselves.[82] Not only did sericulture take advantage of seasonally underemployed labor and household labor but, in particular, it took advantage of female labor within the household. Men could help out, but their primary attention was still devoted to agricultural chores. Traditional gazetteers said that about 90 percent of the labor in sericulture was done by women.[83]

Thus in Kiangnan sericulture was an ideal subsidiary occupation for the peasant household, but not an ideal primary occupation. By not overspecializing, the peasant household could minimize and spread out its risks. A survey from the early 1930s showed that even at Wu-hsing, where sericulture was relatively specialized, if a family had ten mou of land, it would still devote only three or four to mulberry cultivation. At Chia-hsing, the average farm family held 20 mou of land, of which it would devote 3.82 to mulberry. At Wusih, the average family held 10 mou of land, of which it devoted an average of 2.33 mou to mulberry.[84] A typical household in Kiangnan might produce 30–80 catties of fresh cocoons. To produce more than 300 was considered extraordinary.[85] According to another source, the average family output ranged from 32 catties in Fu-yang (Hangchow) to 225 catties in Ch'ang-hsing (Hu-chou).[86]

While sericulture was a secondary occupation, it still played a critical role in the economic survival of the peasant household. Without sericulture, poor families could not pay their debts and taxes.[87] The poor not only earned an income from raising silkworms and reeling silk; they also used the silk wadding in quilts and jackets, and sometimes ate the dead chrysalis like preserved fish.[88] Although mulberry land accounted for no more than 36 percent of farm land, even in the most specialized localities in Kiangnan, a larger proportion of peasant households engaged in sericulture and derived some subsidiary income from it, as Table 24 shows. The survey done in the early 1930s shows that income from sericulture, silk reeling, and weaving formed 42.6 percent of the average peasant household income in Wu-hsing, 24.8 percent in Chia-hsing, and 21.1 percent in Wusih. Even with this income, however, most Kiangnan peasant families found themselves in debt every year, as Table 25 shows. A Japanese study of sericulture in the Wusih area found that silkworms raised by peasant families formed the only cash income they had. Even as late as 1940, they found that the income from the sale of cocoons was far greater than the income they could derive from doing manual labor in the city.[89]

In south China the role played by sericulture in the rural economy differed in important respects. In the Canton delta

TABLE 25 Annual Household Income and Expenditure,
Kiangnan and Kwangtung, 1920s and 1930s
(in yuan)

	Income				
	Total Income	*Silk Income*	*Silk as Percentage of Total*	*Expenditure*	*Difference*
Kiangnan, c. 1931					
Wu-hsing	255.36	108.78	42.6	294.12	−38.76
Chia-hsing	279.14	69.31	24.8	389.63	−110.49
Wusih	365.68	77.27	21.1	415.13	−49.45
Kwangtung, c. 1923					
	939.00	819.00	87.2	601.00	+338.00

Sources: For Kiangnan, D. K. Lieu, pp. 59–67. These figures are based on data gathered about 1931, a period of decline in the silk industry. For Kwangtung, Charles W. Howard and Karl P. Buswell, *A Survey of the Silk Industry of South China* (Hong Kong, 1925), pp. 112–115. These figures represent an estimate, based on field data from about 1923, a period of prosperity for the silk industry. Income from silk is taken to mean income from sericulture, silk reeling, and silk weaving.

districts of Shun-te, Nan-hai, and Hsiang-shan, the farm economy was based primarily on sericulture. In the "4 water and 6 land" system of cultivation, four-tenths of the land was dug out to form ponds, and the mud was used to raise the embankments which formed the other six-tenths of the land. The ponds were used as fisheries while the raised land was used for mulberry cultivation.[90] A field survey conducted by Lingnan Agricultural College around 1923 showed that with seven crops a year, if a peasant household had ten mou of land, of which four were devoted to fish ponds and six to mulberry trees, it could derive an average annual income of 120 yuan from the former and 819 yuan from the latter, forming a total annual income of 939 yuan (see Table 25). Since the annual cost of living for a peasant family of eight people, including food, clothing, and rent, was 601 yuan, the study concluded that the household could claim a net profit of 338 yuan a year. The peasant household in the Canton delta thus earned a substantially higher income from sericulture than its Kiangnan counterpart, although the field surveys cited here were admittedly not conducted in the same period of time.

Although Kwangtung produced and exported as much silk as Chekiang did by the twentieth century, its silk-producing localities were concentrated more narrowly. As Table 26 shows, outside of Shun-te, Nan-hai, and Hsiang-shan, the percentages of the population engaged in sericulture were relatively small. According to the Lingnan Agricultural College's estimate, Kwangtung had the capacity to increase its mulberry acreage about fourfold.[91] Given this capacity, and given the demonstrated profitability of sericulture and related activities, what accounted for the industry's failure to expand further? The Lingnan study stressed reasons similar to those just cited for Kiangnan. First, increased mulberry acreage would further cut down on local rice production, a particularly serious problem in an already rice-deficient area.[92] Second, silkworm disease affected sericulture in Kwangtung as seriously as it did in Kiangnan. Some authorities claimed that a fourfold increase in production could result merely from the elimination of disease; the Lingnan study more conservatively estimated a 50 percent total loss resulting from disease.[93] This in addition to the unstable domestic and international market meant that sericulture in the south was not immune from the serious risks which characterized the enterprise in other parts of China. "There is probably no industry in South China so subject to fluctuation of prices as sericulture," said the Lingnan report.[94] Although sericulture thus formed an essential part of the household economy of peasant families in both Kiangnan and Kwangtung and appeared to have a potential for expansion, closer examination of the household and regional economies suggests why the bureaucratic myth about sericulture was not matched by economic realities in either region.

TABLE 26 Mulberry Acreage and Population in Sericulture
in Kwangtung, c. 1923

Place	Mulberry Land (mou)	Percentage of Total Land
Shun-te		
(Shuntak)	665,000	70
Hsiang-shan		
(Heungshaan)	328,800	—
Nan-hai		
(Naamhoi)	300,000	—
Hsin-hui		
(Sunwui)	60,000	10
San-shui		
(Saamshui)	30,000	10
Pan-yu		
(Poonyue)	10,000	—
Tung-wan & East River		
(Tungkoon)	28,600	—
West River Districts	17,000	—
Ch'ing-yuan & Other		
North River Districts		
(Tsinguen)	16,000	—
Ho-shan		
(Hokshaan)	4,825	—
Ssu-hui		
(Sewui)	5,000	—
Southwestern Districts	500	—
TOTAL	1,465,725	—

TABLE 26 (continued)

Population in Sericulture	Total Population	Percentage of Total Population
1,440,000	1,800,000	80
382,600	—	—
200,000	420,000	48
50,000	1,000,000	5
25,000	320,000	7.8
15,000	—	—
25,000 20,000	2,245,000	1.1
10,000	515,000	1.9
20,000	420,000	4.8
1,000	205,000	0.5
—	—	—
2,188,600	—	—

Source: Howard and Buswell, pp. 15–37. The same material is reproduced in "The Silk Industry in Kwangtung Province," *Chinese Economic Journal* 5.1: 604–620 (July 1929).

RURAL COMMERCE AND FOREIGN TRADE

Although self-sufficient in many respects, the peasant household was tied into a rural commercial network—a network which did not change in its fundamental organization and characteristics with the onset of Western trade. The traditional commercial institutions were flexible enough to deal with the export trade without creating new mechanisms. Some of the basic characteristics of traditional manufacture and commerce, as seen in the Imperial Weaving Factories and elsewhere, were: first, it was extremely specialized in function. Second, it was quite decentralized; although relatively large-scale weaving enterprises existed, one of their principles of operation was the diffusion of production, management, and financial responsibility. Third, the principal means by which responsibility was diffused was through a chain of contractual agreements in which the brokers or middlemen formed the key links. As a result, those who engaged in the silk business did not necessarily have a long-term stake in improving the quality of the product. They simply invested their funds hoping to get a quick and high return.

Figure 5 (in Chapter 2) shows how the domestic silk weaving industry was organized, while Figure 8 (in this chapter) shows how the silk industry was organized for the export trade. The latter is a simplified diagram, which makes no attempt to show local variations, or the many middlemen who might intervene between the basic stages.[95] The organization of rural commerce for export was even more decentralized than that for the domestic weaving industry. The peasant household that engaged in sericulture—whether for domestic or foreign markets—had business transactions with several commercial organizations. The typical household purchased mulberry saplings from the market to start its mulberry trees.[96] It also purchased extra mulberry leaves from the leaf market. In addition, households in the Hangchow and Chia-hsing areas tended to purchase their silkworm eggs from the market rather than starting their own,

FIGURE 8 The Organization of the Silk Export Trade

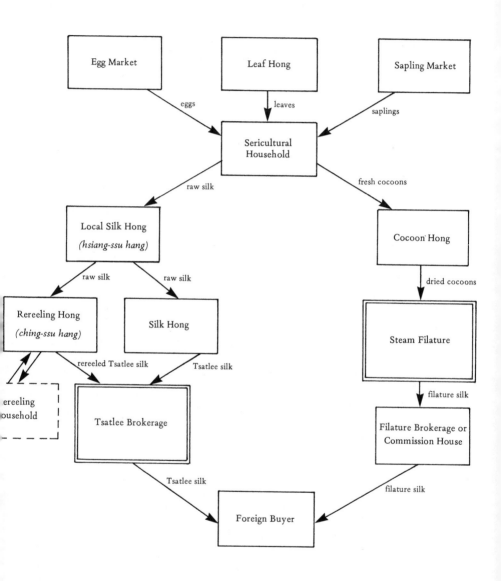

as was customary in Hu-chou and Wusih.[97] One reason they preferred to purchase their eggs is that they were busy reeling the silk from their cocoons just at the time when the moths would be laying their eggs.[98] In the late nineteenth century, Sheng hsien on the T'sao-o River was the major production center, but it was overtaken around the turn of the century by Yü-hang.[99] In the twentieth century, the main centers for the manufacture of egg-sheets in Kiangnan were Yü-hang and Shao-hsing.

At Yü-hang, the manufacture of egg-sheets was a major activity. About 3,000–4,000 households, or about 70 percent of the population, were involved in this line of work. Each year about 300,000 to 500,000 sheets were produced. The opening of the egg market was an annual event of great intensity and excitement. Around May twenty-first, about a thousand merchants would come from other localities in Kiangnan, primarily from northern Chekiang, to purchase egg-sheets.[100] Those eggs, usually of inferior quality, which were not marketed immediately in this way, were shipped to the Anhwei and northern Kiangsu areas and sold at a much lower price.[101]

Because the earliest egg-sheets to appear on the market would bring the higher prices, the competition to get one's eggs on the market was fierce. The manufacturers of the egg-sheets would often resort to underhanded methods, such as using heat to hasten their maturation.[102] The local guild of egg merchants tried to control the competition by forbidding the sale of eggs before an officially established date. The guild also had a common price scale—the standard for which was set by the most prestigious dealer in Yü-hang, Wu Fu-hsiang (Woo Foo-tsing). His eggs were considered so far superior to anyone else's that the peasants would often pay large deposits as much as six months in advance for his egg-sheets.[103]

Except for purchasing saplings, extra leaves, and possibly egg-sheets, the typical peasant household was self-sufficient as far as the materials and equipment needed for raising silkworms and reeling silk. If producing for the domestic market, the household would sell its reeled silk to a silk hong through the medium of a local or itinerant merchant. As seen in Figure 5, the silk hong, in turn, would supply an account house or independent weavers with raw silk. Some peasant households would spin or twist the silk into warp before selling it to a warp hong. At Chen-tse and other places, this was a major peasant occupation. This type of warp was called *hsiang-ching,* or native or local warp, while warp which was made from raw silk purchased from a hong was called *liao-ching.* The latter was sold to various dealers from the silk weaving centers. Warp was further subdivided into specialties such as *Su-ching,* which was warp sold to Soochow hongs, or *Kuang-ching,* which was warp sold to Canton merchants. In addition there later was *Tung-yang ching,* or warp of Japanese style, twisted from right to left, rather than the Soochow method of left to right, which was adopted in the late nineteenth century. There was also *yang-ching,* various types of warp generally twisted from local silk shipped first to Shanghai.[104]

After Shanghai was opened to foreign trade, there was a dramatic shift in the distribution of key markets and trading routes for Kiangnan silk. At the outset, silk was still shipped to Canton for export abroad.[105] During the Tao-kuang and Hsien-feng periods, Shuang-lin silk merchants prospered in this silk trade, but as silk began to be shipped directly to Shanghai, the Shuang-lin merchants were gradually overshadowed by Nan-hsun and Chen-tse dealers. By the late Ch'ing, very little of the silk export business still originated at Shuang-lin.[106] At Ch'ang-hsing, a similar result occurred after the export trade expanded. The local gazetteer observed that in the past Ch'ang-hsing had at least a hundred silk hongs, but by the end of the century, only ten or so were left. The majority of peasants sold their silk at Nan-hsun or other places, and only 20–30 percent of the Ch'ang-hsing silk was sold locally.[107]

Nan-hsun became the center of the raw silk export trade, and more and more business became concentrated there. The local gazetteer estimated that one-half of the local silk was sold to foreigners.[108] A Japanese survey estimated that, in the 1920s about one-half the silk sold at Nan-hsun was of local origin, while the other half, mostly lower quality silk, came from Shuang-lin, Ling-hu, and other centers. About 20 percent of the silk collected at Nan-hsun was reshipped to Chen-tse, 10 percent was supplied to the weaving districts, and the rest was rereeled for export. Chen-tse, like Nan-hsun, became an important distribution center associated with the export trade. There were about a hundred silk hongs there, but all very small in scale. Eighty percent of their supply came from the hongs of surrounding localities, and most of it was rereeled and shipped to Shanghai for export. A small portion was sent to weavers in Soochow or Canton.[109] Other localities became dependent on the main centers. At Wu-ch'ing, for example, there had never been any warp hongs. All the fine silk produced there was handled by the Chen-tse hongs, whose representatives went to Wu-ch'ing to buy silk to distribute to weavers, while coarse silk continued to be sold to merchants from Hangchow, Shao-hsing, and Sheng-tse.[110]

There were many varieties and sizes of silk hongs. The *Kuang-hang* were the Cantonese merchants who were involved in the early phases of the export trade, but who were later replaced by Hu-chou merchants.[111] Generally speaking, as shown in Figure 8, the peasant household would sell its silk to a local hong, often called a *hsiang-ssu hang* (lit., "local silk hong"), often through the medium of a small merchant called a *hsiao ling-t'ou* (lit., "small remnant"). The *hsiang-ssu hang* were usually small in scale, required very little capital, and were owned by an individual or a partnership. They were regulated by the local silk guilds, who tended, in the opinion of some observers, to force the price below what its free-market level would have been. A *hsiang-ssu hang* would sell its silk to a larger silk hong and, in fact, it was usually given orders by it for the season.[112]

The silk hongs themselves were usually much larger in scale of operations than the *hsiang-ssu hang*. They usually were well capitalized and financed individually or as a joint-stock company.[113] The larger hongs were associated with silk warehouses, or "godowns" at Shanghai to whom they sold their silk. Each had its own chop.[114] There was, of course, a difference between the silk hongs that handled silk for export, and those that supplied silk to the domestic weaving industry and generally operated on a much smaller scale.

Until the 1880s, the silk hongs simply shipped raw silk directly to Shanghai for export but, by the turn of the century, it became the custom for the *hsiang-ssu hang* to sell their silk to the *ching-ssu hang,* which had the silk rereeled before exporting it. It was estimated that there were 2,000–3,000 households in the Nan-hsun area which were engaged in rereeling. Each had an average of four rereeling machines, each of which could rereel an average of ten *liang* of silk per day. Approximately 106 or 107 catties of raw silk were needed to yield 100 catties of rereeled silk. Nan-hsun had about seven or eight hongs which produced about 4,000 piculs of these Tsatlee rereels, while Chen-tse had about twenty or more hongs, which produced 10,000 piculs annually. By the 1920s virtually all of the Tsatlee raw silk which was exported passed through the hands of the *ching-ssu hang,* who ordered silk from the smaller hongs, had the silk rereeled, affixed their trademarks, and sold the silk at Shanghai.[115]

The silk hongs in the interior operated on orders from the brokers in Shanghai to whom they were attached. The relationship between the Tsatlee brokers and the silk hongs was usually close and long-term. In the 1920s, the commission paid to the hong for each hundred yuan's worth of silk was 1.6 yuan.[116] At Shanghai, there were about ten silk brokerages associated with the Tsatlee trade. They were well capitalized, and each had its warehouse, as well as lodgings for visiting merchants. Chen-ch'ang and Chen-t'ai were the largest, each doing 7,000–8,000 piculs of business a year. Their commission was 2.5 taels per bale.[117] These Tsatlee merchants were organized into the Shanghai Silk Merchants' Guild, which in 1924 changed its name to the Kiangsu and Chekiang Silk and Reeled Silk Merchants' Association.[118] There were other silk brokerages at Shanghai for handling silks from other parts of China, as well as those for filature silks. (See Chapter 3 above.)

It was not only the Nan-hsun merchants who got rich quickly in the silk trade. At Shuang-lin, the gazetteer observed that local merchants often went to Soochow, Hangchow, Fukien, or Kwangtung to do business, but those who remained behind could make an easy profit in the silk business. "In the morning you have no food to eat, but by night you have a horse to ride." The gazetteer, however, also noted that the competition in the silk business was extremely keen. People counted on the good markets and did not prepare for hard times. Some invested their money unwisely. Others simply gambled away the wealth of their fathers and grandfathers in a profligate way of life. It was rare to see a family that had stayed wealthy for as many as three generations.[119]

The extremely seasonal nature of sericulture in Kiangnan placed the peasants at a particular disadvantage with respect to the wealthy silk merchants when it came to credit. Peasants often had to borrow money in order to pay for leaves and other expenses. Traditionally it was said that on every thousand *ch'ien* borrowed, the interest due at the end of the season was a hundred *ch'ien*. In other words, the rate of interest during the silkworm season was 10 percent. This was colloquially known as "adding one *ch'ien*."[120] In Ch'ang-hsing it was reported that although 10 percent had previously been the interest rate on silk loans, by the T'ung-chih period 20 percent was charged. The Ch'ang-hsing gazetteer said that "the poor people lose sleep and don't eat in order to raise their silkworms and make silk. Their bitterness is unspeakable. They depend on this for paying their taxes, rents, clothes, food, and daily needs. The city brokers come in their many boats and insult the peasants. This is something that must be endured." The peasants called these silk merchants "the silk devils."[121] The bad reputation of the silk merchants among the peasants was not confined to one locality.[122] At Nan-hsun it was said that there were 110 rich silk families. The largest of them were called "the four elephants, the eight cows, and the seventy-two dogs." The four "elephants" —the Liu, Chang, Pang, Hsing, and Yueh, actually five in number—were said to be worth more than a million taels each, while the eight "cows" were worth half a million or more, and the seventy-two "dogs" were those families worth 300,000 or more.[123]

Although the silk export trade improved the fortunes of many people—peasants and merchants alike—it also exposed them to greater risks. In order to minimize these risks, localities as a whole, and peasant households individually, limited their output of silk. Merchants for their part tried to minimize their risks in a business that was notoriously speculative by maintaining the protective bonds (*pao*) offered by the traditional commercial practices. As observed earlier, however, the Chinese internal marketing mechanism possessed in its very strength a definite weakness. Its strength lay in its flexibility and its ability to diffuse risk. Its weakness, however, was its preoccupation with immediate profits rather than the long-term development of the trade. As C. F. Remer observed, "Chinese marketing of products in general tends to be a succession of commission transactions, of dealings like those of a broker. Chinese merchant guilds bring together those who handle a product at some particular stage. There has been little organization for the purpose of controlling a particular product on its way from producer to consumer."[124]

With the inception of large-scale export trade, the silk industry of Kiangnan in effect split into two sectors—one which serviced the export trade, and the other which continued to supply raw silk to the weaving industry. Coexistence between the two sectors was possible because, at least initially, the export trade did not necessitate a reorganization of rural silk commerce. In the early stage, when there was a considerable expansion of silk output without a change in technology, the network of traditional commercial transactions was simply lengthened by one or two links and redirected toward Shanghai. Only one end of the chain or network was affected; the other remained structurally untouched. Only after technological change was finally introduced in the form of steam filatures were severe tensions created between the export sector and the traditional sector. Until then the flexibility of the traditional risk-diffusing mechanisms developed in rural agriculture and commerce permitted the limited growth of sericulture in response to foreign markets, but it prevented growth sufficient to create structural or institutional changes.

SIX

Foreign Trade and Modern Enterprise

Many foreigners in the silk trade in the late nineteenth century saw industrialization as a simple solution for the reform of the Chinese silk industry. "Had they the simplest of filatures, or even if they adopted the system of drying their cocoons, so as to be able to reel more slowly and carefully, they could increase the value of their out-turns some 30 percent," complained one foreign observer.[1] Although the steam filature industry was established in Kwangtung by the 1870s, in Shanghai it was not until the 1890s that mechanized factory reeling took root. Despite this slow start, however, silk reeling was among China's most important modern industries in the twentieth century, representing 97 out of a total of 549 Chinese-owned enterprises started between 1895 and 1913.[2] Like most modern industries in China, however, silk reeling suffered from inadequate financing and inadequate, or at least inappropriate, government support.

In addition the filature industry had difficulty securing the supply of high-quality cocoons that it needed, a problem which stemmed in part from its poor integration with its rural base and which it possessed to a unique degree among all the modern industries.

The earliest attempts to establish steam filatures at Shanghai and Canton during the 1860s were conspicuous failures. The first known filature at Shanghai, sponsored by Jardine, Matheson, and Company in 1862, failed allegedly because Chinese officials and merchants, anxious to protect traditional interests, prevented it from procuring an adequate supply of cocoons. This filature closed within four years.[3] In Canton in 1866, the pioneering effort of Ch'en Ch'i-yuan (Chan Kai-uen), who had studied reeling techniques in France, met with opposition from local handicraft workers and also peasants who thought his facory to be inauspicious or threatening. Although the factory was dismantled and reassembled in Macao, within three years Ch'en was asked to return to the Canton area, and by 1881 there were ten modern filatures with 2,400 basins in Kwangtung province.[4]

Thereafter the steam filature industry grew rapidly in the Canton delta, mostly in Shun-te and Nan-hai hsiens. By 1895, over 85 percent of Canton's raw silk exports consisted of filature silk, and by the turn of the century virtually all exports were steam filatures (see Table 12). By 1912, there were 109 filatures with 42,100 basins; by 1918, 147 with 72,200 basins; by 1922, 180 filatures with 90,064 basins; and by 1926, 202 filatures with 95,215 basins.[5] Although Canton filatures were originally rather small in scale, eventually they expanded. The majority had between 400 and 500 basins.[6]

The first successful filature at Shanghai was the Keechong Company (Ch'i-ch'ang ssu-ch'ang), which was sponsored by the American firm of Russell and Company, and had a Frenchman, Paul Brunat, as manager. This company later expanded and changed its name to the Shanghai Silk Filature (Pao-ch'ang ssu-ch'ang) and had as many as 1,000 reels.[7] A second attempt by Jardine, Matheson, and Company in 1882 met with greater success than its first had. The Shanghai Silk Manufacturing Company (Shanghai ts'ao-ssu-ch'ang), later renamed the Ewo Filature (I-ho ssu-ch'ang), was financed 40 percent by foreign capital and 60 percent by Chinese capital. It had 500 reels and employed over 1,000 workers.[8] At about the same time a Hu-chou silk merchant named Huang Tso-ch'ing, known to Westerners as Ch'ang-chi, financed the establishment of Iverson and Company (Kung-ho yung), which, together with another new filature, the Shanghai Filature Company, started operations in 1882 under the direction of an Italian engineer. All of these filatures experienced considerable difficulties initially, including the lack of skilled workers and marketing problems.[9] In the 1890s these filatures were joined by several others, including one English filature of about 200 reels, a French filature of 530 reels, and a German concern of about 480 reels.[10] By 1901, there were 23–28 filatures in Shanghai with 7,800–7,900 basins in total, an average of 278 to 340 basins per filature, as shown in Table 27.[11]

In the first decade of the twentieth century, the number of filatures at Shanghai increased only slightly. Between 1908 and 1911, however, the number increased sharply from 29 to 48, reaching a total of 13,738 basins. After 1912, the growth of filatures continued steadily until about 1920, when the effects of postwar recession were felt. During the mid-1920s, the number of filatures again climbed steadily, reaching a peak in 1930, when there were 107 filatures with a total of 25,395 basins in operation.[12] It should be noted, however, that the number of filatures actually in operation in any given year varied a great deal, which is why statistics often seem inconsistent.

Initially Shanghai offered many advantages as a location for steam filatures—extraterritoriality, easy financing, access to foreign capital, convenience for shipping, and cheap labor. But as early as the turn of the century, these advantages seemed to be outweighed by new disadvantages. Competition for cocoons increased with the growing number of filatures in Shanghai. Real estate values soared. Labor became progressively more

TABLE 27 Steam Filatures in Shanghai, 1890–1936

Year	No. Filatures	No. Basins	Year	No. Filatures	No. Basins
1890	5	—	1914	56	14,424
1891	5	—	1915	56	14,424
1892	8	—	1916	61	16,692
1893	9	—	1917	70	18,386
1894	19	—	1918	68	18,800
1895	12	—	1919	65	18,306
1896	17	—	1920	63	18,146
1897	25	7,500	1921	58	15,770
1898	24	7,700	1922	65	17,260
1899	17	5,800	1923	74	18,546
1900	18	5,900	1924	72	17,554
1901	23	7,830	1925	75	18,298
1902	21	7,306	1926	81	18,664
1903	24	8,526	1927	93	22,168
1904	22	7,826	1928	104	23,911
1905	22	7,610	1929	104	24,423
1906	23	8,026	1930	107	25,395
1907	28	9,686	1931	70	18,326
1908	29	10,006	1932	46	12,262
1909	35	11,085	1933	61	15,016
1910	46	13,298	1934	44	—
1911	48	13,738	1935	33	7,686
1912	48	13,392	1936	49	11,094
1913	49	13,392			

Source: Yen Chung-p'ing, et al., comp, *Chung-kuo chin-tai ching-chi shih t'ung-chi tzu-liao hsuan-chi* (Peking, 1955), pp. 162-163. The data for 1890–1932 are also found in H. D. Fong, p. 494.

expensive.[13] Moreover, without foreign support, it was difficult for a Chinese to establish a filature in Shanghai, and as a result, some filatures were gradually established in the interior.[14] Of the various other places where filatures were established, only

Wusih experienced considerable success. The first filature, the Yü-ch'ang, was established there around 1904.[15] Between 1909 and 1910, four other filatures were built, but the silk reeling industry did not assume great importance at Wusih until after World War I. During the 1920s, the number of filatures increased rapidly. In 1925, there were eighteen or nineteen filatures; in 1930, there were forty-two.[16] Not only were wages at Wusih lower than those at Shanghai, but labor unrest in Shanghai during the 1920s made the interior an even more attractive place to start new factories.[17]

The steam filature industry in Wusih served as a direct stimulus to the growth of sericulture in that area. Proximity to a supply of cocoons in turn conferred several advantages. First, there was some savings in the amount of likin and transit tax that had to be paid. Cocoons shipped to Shanghai for reeling paid about 55.39 taels total for cocoons to yield 100 catties of silk. Silk reeled first at Wusih and then shipped to Shanghai was levied a total of about 38.46 taels per 100 catties.[18] The proximity to sericultural areas meant that Wusih filatures could receive their cocoons a half a month to a month earlier than Shanghai filatures. This advantage of time was particularly critical in an industry that was so competitive and so speculative. Finally, the cost of shipping raw silk to Shanghai for export was less than that of shipping cocoons there for reeling. According to one estimate, all these advantages together meant that Wusih silk was thirty taels per picul cheaper than Shanghai filature silk.[19]

In the old silk districts, however, the advantages of cheap rail transportation after the construction of the Shanghai-Ningpo-Hangchow rail line were more than offset by the resistance of the traditional weaving interests to the introduction of filatures. In 1912, the provincial government of Chekiang set up five model filatures in Hai-ning, Ch'ung-te, Sheng hsien, Te-ch'ing, and Wu-hsing, using Japanese-style reeling machines, but these did not seem to have made much of an impact.[20] At Te-ch'ing, a further attempt to set up a model filature in 1914, with foot-powered machines and rereeling machines, ended in failure by 1916.[21] Only Hangchow was an exception to the general pattern of nonreceptivity to filatures in the old districts. Although the early attempts to open filatures around 1895 were isolated cases not forming a general trend, during the 1912–1919 period, the modern weaving factories that were established in Hangchow created a demand for a convenient source of filature silk.[22] Hsu Ping-k'un, head of a local technical school, set up a machine-training center, and imported machines and instructors from Japan. Three filatures were established in 1912, and by 1919, there were nine filatures in Hangchow.[23] In 1925 all but one of the five filatures still operating in Hangchow were owned by weaving companies. All were considered very successful.[24]

At Chia-hsing, a branch of the Wei-ch'eng Company of Hangchow was opened in the 1920s, and two or three other filatures were also started.[25] In addition, during the 1920s and early 1930s, a handful of filatures were opened at Wu-hsing, Te-ch'ing, Hai-ning, and Hsiao-shan.[26] At Soochow, with the exception of the Su-ching Company, and two other joint official-merchant enterprises, conservative attitudes prevailed and steam filatures never caught on. The success of the Su-ching Company was in large measure due to the influence of Wang Chia-liu, a local notable who had acquired control of it. Wang also owned two of the local *ch'ien-chuang* (Chinese native banks), which financed the purchase of fresh cocoons for the filatures. The manager of the Su-ching, Wang Ts'un-chih, was also well-connected, and managed the Heng-li filature as well as owning two cocoon hongs at Wusih and one at Soochow.[27]

Between 1914 and 1928, the average size of filatures in the Kiangnan area was 262 basins per filature, as shown in Table 28. This was relatively small compared to those in Canton. In 1928, out of a total 160 or 162 filatures, only 2 had fewer than 100 basins, 34 had more than 100, 91 had more than 200, 16 had more than 300, 9 had more than 400, 6 had more than 500, and 2 had more than 600.[28]

FINANCE AND LABOR

Despite the immediate popularity of steam filature silks, and their domination of the raw silk market after 1895, the steam filature business was full of hazards for entrepreneurs. Many filatures closed down after a brief period of operation. Most of the early filatures that survived were undercapitalized and constantly on the verge of bankruptcy.[29] Although the filatures at Shanghai were initially larger than those at Canton, in 1913 the average investment for all was only 119,000 yuan, or about 79,000 taels, according to one estimate.[30] According to another estimate, around 1904 it took about 70,000 taels to set up a filature at Shanghai: 40,000 for construction of the factory, 12,000 for the reeling machines, 8,000 or more for steam engines, and 10,000 for miscellaneous equipment.[31] In general, it was extremely rare to have a filature with more than 200,000 taels capital investment.[32] The average capital investment in later decades ranged from about 35,000 to 100,000 yuan, or between approximately 23,000 to 67,000 taels.[33]

The scarcity of capital for steam filatures did not escape the attention of provincial leaders such as Li Hung-chang and Chang Chih-tung. In 1883, Li Hung-chang, then governor-general of Chihli, took up the problem of the likin on cocoons with the British minister. Li observed that the equipment and machinery of the filatures that had failed in Shanghai were about to be shipped off to Japan.[34] In 1887, he reported on his tour of the Keechong and Jardine filatures at Shanghai and expressed his admiration for the silk that they manufactured. He suggested that the governor-general and governor of Chekiang be ordered to solicit shares from merchants in the silk districts in order to establish filatures in places like Chia-hsing and Hu-chou. He noted that only through enlisting merchant support would such experiments succeed; enterprises managed solely by officials would probably fail.[35]

In the 1894–1896 period, Chang Chih-tung, as governor-general of Liang-Kiang, expressed a similar interest in joint official-

TABLE 28 Steam Filatures in the Kiangnan Area, 1914–1928

Year	Shanghai	Wusih	Others	Total Filatures	Total Basins	Average No. Basins Per Filature
1914	60	5	8	73	19,208	263
1915	57	6	8	71	18,996	268
1916	59	8	8	75	20,480	273
1917	70	8	8	87	22,794	262
1918	68	9	8	85	23,336	275
1919	65	14	10	89	24,406	274
1920	63	14	10	87	22,246	256
1921	58	15	10	83	22,202	267
1922	65	14	11	90	23,752	264
1923	74	18	12	104	26,782	258
1924	72	18	11	101	25,618	254
1925	75	17	12	104	26,182	252
1926	81	20	16	117	28,697	245
1927	88	22	18	128	33,660	263
1928	95	38	29	162	40,354	249
					average:	262

Source: *Shina sanshigyō taikan,* pp. 235–236. Corrections made for typographical errors in the original. Discrepancies between these figures and those in Table 27 can be attributed to errors in determining which filatures were actually in operation in given years.

merchant enterprises. He noted that local merchants were reluctant to invest their money in modern enterprises such as silk reeling, cotton spinning, or matchmaking factories, without some government support. He proposed borrowing over two million taels to establish a commercial bureau to support such projects. The Su-ching Filature and the Su-lun Cotton Mill were two results of this effort.[36] In subsequent years, the official shares were sold, and control of the companies fell to Wang Chia-liu.[37]

Despite these efforts, the silk reeling industry remained largely free from both bureaucratic and foreign direct investment. In fact its development in Kwangtung prior to 1895 was a conspicuous example of early Chinese investment in modern enterprises.[38] At Shanghai, foreign investments, which had played a leading role initially, were much less important by the turn of the century, when most foreign filatures came under Chinese control.[39] In 1911, only five of Shanghai's filatures were owned by Europeans.[40] Western ownership was often nominal since Chinese owners and investors would use Western names and brands as fronts, partly to secure extraterritorial protection for the treaty port enterprises, and partly to gain the advantage of using Western chops which were familiar in foreign markets. Often as compradores and officials began to take over the ownership of a filature, they would try to conceal the fact and maintain the facade of Western ownership.[41] For the most part, however, direct foreign investment in the filature business in the twentieth century was minimal.[42]

The single greatest exception to this generalization was the Japanese, who played a significant role in the Chinese silk industry. As early as 1906, the Shanghai Silk Spinning Company (Shanghai chih-tsao chüan-ssu ku-fen kung-ssu) was established as a joint Sino-Japanese enterprise for the processing of waste silk. The chairman of its board was Chu Pao-san, a leading figure in Shanghai's self-government movement.[43] Later, Japanese had interests in businesses at Shanghai, Wusih, Hankow, Szechwan, and Manchuria.[44] In the 1920s and 1930s, the Japanese tried to increase the output of Chinese cocoons for the Japanese silk industry, for which they were often willing to pay quite high prices, thus making things even more difficult for their Chinese filature competitors.[45]

In the absence of Western or bureaucratic capital, the silk reeling industry resorted to several methods of stretching a limited amount of private Chinese capital. At Shanghai an outstanding characteristic of the filature industry was that the filature operators and the factory owners were not the same people.

When the filaturists or entrepreneurs wished to go into business, they simply rented the building and equipment from the owners —and sometimes the trademark or chop of the filature as well.[46] About 80–90 percent of the filatures in Shanghai were rented rather than being owned by their operators. Usually the rental was done through a yearly contract, at the rate of about two to four taels per basin per month.[47] After World War I, the rate went up to three taels, and by the 1930s, it was over three taels.[48] It was estimated that in a good year the filature owner could receive a 15 percent return on his property with no difficulty.[49] At Wusih, most filatures except those controlled by the Hsueh family were also rented in this manner.[50] In the opinion of the Shanghai Testing House's survey, however, one of the great advantages that Wusih's filatures had over Shanghai's was the "ample financial backing by local capitalists, which makes them sound and successful enterprises, [which avoid] continual changing of management and organization."[51]

Almost all the Shanghai filatures were operated as partnerships. The partners put up the bulk of the capital, and the *ch'ien-chuang,* or other sources, supplied the rest.[52] Usually one or two of the original partners was chosen to be the manager, and he had the ultimate and direct responsibility for the enterprise.[53] The *ch'ien-chuang* played a central role in the filature business. Not only did they finance the purchase of cocoons but, because most of the silk business was conducted on a cash basis, their funding was absolutely critical to the operation of the whole system. In effect, as the Japanese observed, the *ch'ien-chuang* served as middlemen between the filaturists and the capitalists. And because of the degree to which they financed the filature's operations, a person with very little capital could engage in a rather large-scale enterprise.[54]

The advantage of this system was that it minimized the risk to all parties. For the capitalist, the filature was a relatively safe real estate investment. For the operator or entrepreneur, there was no long-term commitment or risk. Business was done only when the season promised to be profitable. The entrepreneur needed to use his capital only on operating costs and the purchase of cocoons. The future of the physical plant was not part of his concern, and in this way he could move in and out of the silk reeling business with relative ease.[55] For the silk business as a whole, however, such methods of financing had extremely unfortunate consequences and contributed to its speculative nature. As one Chinese article said, "From the point of view of the lender [of the money], it is a form of speculation, and from the point of view of the borrower, it is a form of usury."[56] As the Shanghai Testing House survey of 1925 observed, "The operation of the filatures on such a small capital, buying from hand to mouth, makes the business very unstable." If some financial impasse was reached, the filature would simply change management, and the constant change in management meant that the "chops" did not have reliable reputations.[57] The fluctuating number of filatures in operation was a manifestation of this instability.[58] And since the filature operators had no investment in the plant and equipment, there was no incentive to modernize or reform.[59] In the 1922–1931 *Decennial Report,* the Shanghai Customs Commissioner observed that most of the filatures in Shanghai had machinery that was thirty years old, and there was only one filature which had "thoroughly up-to-date" equipment.[60]

The financing of filatures in the Canton delta was no more stable than in Shanghai. A partnership, usually of relatives, was the typical form of business. The capitalization of filatures ranged from 30,000–200,000 yuan (roughly 20,000–133,000 taels) and averaged about 100,000 yuan.[61] Filatures were usually built by a clan or family and then rented to an operating company, in a manner similar to that of Shanghai filatures. Financing was perhaps even more precarious than at Shanghai. Some stockholders of filature companies guarded against losses by investing in several small companies. Companies frequently changed personnel and even names. Consequently the Canton filatures were not kept in very good condition; by the twentieth century their machinery and equipment were already considered to be very backward.[62] A Western observer later remarked, "A Cantonese filature looks as if it has been in existence for centuries, the only thing recalling present times is the steam engine. Dirty walls, dark rooms, stingy oil lamps hanging from the ceiling, —it is a sight that at once recalls the conservative people of South China. How different are the Chinese people of the north."[63]

Canton's labor force, however, was organized on a more stable basis than Shanghai's, since the filatures could operate all year around. The *tzu-shu nü* (lit., "women who comb their own hair"), an unusual group of Cantonese women who took vows in a sisterhood, formed an important part of the filature work force.[64] At Shanghai, the casualness of labor arrangements was a problem for the filatures. Complaints about the workers' lack of skill and discipline were frequently voiced. One observer said that, compared to the orderliness of Japanese filatures, the Chinese factories were utterly chaotic. Workers were lazy, sloppy, dirty (they combed their hair in the reeling room and boiled corn ears in the cocoon basins!), and dishonest.[65] While filature workers may not have been instinctively any more lazy or dishonest than workers in other industries, there were certain conditions which promoted less-than-perfect standards. In the first place, the filature labor force was a shifting one. Since the Shanghai filatures operated on a seasonal basis and some did not even operate every season, a filature did not have a fixed labor force. Instead each would recruit its work force on an ad hoc basis from the pool of thousands of filature workers in Shanghai or Wusih. No particular experience was required, and often a female worker would simply bring a relative along.[66]

Most of the workers in the filature industry were women, but the foremen and the engine room workers were men. The majority of women were reelers—regular reelers, substitute reelers to take their place for short periods, and basin workers, young girls who managed the cocoon-boiling basins. Women also did cocoon peeling, a very menial sort of work done on a piece-rate basis. In total, it was estimated that about 245–256 workers were needed for every 100 reels. Thus a typical filature of 240 reels would employ over 500 workers.[67] Wages for the reelers were calculated on a time basis within guidelines established by the Cocoon and Steam Filature Guild. Some filatures used bonuses as an incentive to encourage the workers to get better yields and imposed penalties if their work was not up to standard.[68] The usual working day was eleven or twelve hours, with one hour or more for lunch and breaks. At Shanghai, usually one day of rest was given every two weeks. There was no contract, and at Shanghai there was a custom by which the foreman retained the worker's first two weeks' wages, in theory, as a bond or deposit against the worker's early resignation but, in fact, as a form of payment to the foreman.[69]

Another unhappy aspect of the labor situation in filatures was the use and abuse of child labor. The basins of steaming water which were used to loosen the strands of silk from the dried and peeled cocoons were managed by children sometimes as young as eight or nine years old.[70] An investigation conducted by the Bureau of Social Affairs of the Shanghai Municipal Government in the 1930s found that the conditions in many filatures were appalling: "Tiny children stood for an eleven-hour day, soaked to the skin in a steamy atmosphere hot even in winter, their fingers blanched to the knuckles and their little bodies swaying from one tired foot to the other, kept at their task by a firm overseer who did not hesitate to beat those whose attention wandered."[71] In Europe and Japan these tasks were performed by adults.[72] Only after 1937, when the filature industry at Shanghai was revived and new filatures established in the Shanghai International Settlement, were changes made: a central boiling system placed the basins outdoors and away from the rest of the workers, and a slow-reeling system made it unnecessary to beat the cocoons in the basins; no children had to be employed.[73]

Although there is no direct evidence of it in silk filatures, silk weaving factories at Shanghai employed labor contractors, who controlled several subcontractors. The chief contractor would distribute the total amount of wages paid by the mill, minus a commission for himself, to the subcontractor, who in turn had the same relationship with the sub-subcontractor, who in turn paid the wages to the workers.

> Thus the wages received by the workers directly engaged in production go through many hands and sustain many deductions. . . . Under this system of contract labor no employer-employee relationship exists between the workers and the mill. . . . The advantage to the mill under this system seems to be its escape from the troublesome problem of labor management.[74]

This *pao-kung* (lit., "guarantee work") system was also widely employed in the Shanghai cotton mills, where it also saved Westerners unfamiliar with local practices the trouble of having to hire their own labor force.[75] It could be considered a twentieth-century version of the *pao-lan* or other contractual arrangements used by the Imperial Silk Weaving Factories.

COCOON HONGS AND MARKETS

The second greatest problem faced by steam filatures, particularly at Shanghai, was that of finding an adequate supply of good cocoons. Since cocoons represented 75–80 percent of the cost of producing raw silk, their amount, price, and quality were of critical importance to a filature's success.[76] In Kiangnan, cocoon hongs were established at rural centers to buy cocoons from the local peasants, dry them in ovens, and then ship them to Shanghai for storage and eventual reeling. The cocoon hong itself was a building in which a number of ovens for drying the cocoons were located.[77] Like the other types of hongs in the silk trade, the cocoon hong operated only with an official license and generally had the monopoly of business in its locality. In some respects the cocoon hong appeared to be yet another link in the chain of brokers and middlemen in the silk business. Unlike the silk hong, however, the cocoon hong was engaged in the processing of a particular raw material on its premises and not simply in a commission transaction.[78] And unlike the silk hong, the cocoon hong had a direct relationship with the modernized sector.

The cocoon hongs were usually owned by local people of wealth.[79] Establishing a cocoon hong required 20,000–30,000 yuan (roughly 13,000–20,000 taels) on the average, although a large one could cost as much as 60,000–70,000 yuan (40,000–46,600 taels), and a small one could be established for about 10,000 yuan.[80] Most commonly, the hongs were owned either by an individual or by a group of friends in a partnership arrangement. Very rarely was the cocoon hong financed as a company.[81] In 1930, for example, the province of Kiangsu had 1,318 cocoon hongs, of which 352 were individually owned, 830 were jointly owned, 2 were owned by a company, and 134 were financed in an unknown manner.[82] According to one estimate, only about one-tenth of the cocoon hongs were owned by filatures.[83] In the early days there were cases where a cocoon hong was built on the premises of a filature, but this practice was almost completely discontinued.[84]

A steam filature established a relationship with a cocoon hong either by renting or commissioning it. There was much local variation in these arrangements, with varying degrees of responsibility assumed by the filature and by the hong. The most common method was outright rental, or *tsu-tsao*. All the facilities and equipment were rented to the filature for a period of a year or two. The rental averaged between 500 and 800 yuan, but in Hai-ning and Chia-hsing, where there were relatively few hongs before the mid-1920s, the rental could reach 2,000–3,000 yuan or more. Generally, the charge was about 50–100 yuan per oven.[85] The filature assumed all the responsibility for operating the hong, including purchasing and processing the cocoons.[86] On occasion, the hong might be subleased to a third party.[87]

The second method, called *pao-hung* (lit., "contract drying"), involved contracting with the hong for the drying of cocoons which the renter would buy. Payment was made for this service according to the amount of the cocoons dried. In Wusih the charge was twelve yuan per picul, while in Li-yang and I-hsing, it was eighteen yuan per picul.[88] The hong took the responsibility for hiring the help and operating the equipment. This method was used in Wusih and other places in Kiangsu. In the third method, called *pao-chiao* (lit., "contract delivery"), the same procedure was followed except that the contract provided that the hong assume the responsibility for shipping the cocoons to Shanghai, and the fee included the cost of shipment. This was practiced at Shao-hsing and Anhwei most commonly. Since the second and third methods differed only by degree, the one was called *hsiao-pao,* or "little contract," while the other was called *ta-pao,* or "big contract."[89] The advantage of these two methods was that the renter did not need to be familiar with the local area since the hong took the responsibility for the management and operation of the hong, including the hiring of personnel.[90] In both cases, the renter assumed the name or trademark of the hong as part of the contract.

The fourth method, called *pao-shou* (lit., "contract for receipt"), involved the least amount of direct responsibility on the part of the renter. The hong purchased and processed the cocoons on orders from a filature. There were many variations on this pattern. This system was used in places where transportation was inconvenient and political and military conditions were not safe, or where political and military conditions disrupted normal transport routes. The commission could be calculated and paid in numerous ways.[91]

Only one portion of the cocoons processed by the cocoon hongs during the cocoon season was directly purchased or ordered by the filatures. The rest was handled through cocoon brokers who purchased, dried, and stored a supply of cocoons to sell to the filatures later in the year to supplement their needs. In Chinese these brokers were called *yü-chien shang* (lit., "surplus cocoon merchants"). Occasionally, these brokers owned their own hongs, but more often they made an arrangement with the hong in one of the four ways mentioned above.

The brokers dealt in cocoons on a speculative basis, hoping that the demand and prices for cocoons would rise during the year. They also created a kind of buffer for the filatures so that the latter did not need to risk purchasing a larger supply of cocoons than it might need during the year.[92] Estimates of the portion of the cocoon crop handled in this way vary. According to most, the amount purchased directly by the filatures each season varied from 30–80 percent of their total needs for the season; the rest was purchased later.[93] On the other hand, the number of hongs operated or commissioned by the filatures was about 60 percent of the total, while the number operated or commissioned by the brokers was only 40 percent. In any case by 1932, only 10–20 percent of the hongs were engaged by the filatures, 20–30 percent by the brokers, and the rest had closed down.[94] It appears that as the silk business declined, filatures became increasingly reluctant to commit their funds to a large quantity of cocoons.

The cocoon business may have been the most speculative sector of the silk trade, since it operated under the most severe time constraint. Cocoon prices were subject to wide annual fluctuations.[95] The cocoon purchasing season was compressed into a period of two to seven days.[96] In the Kiangnan area, all the cocoon markets—from the earliest at Shao-hsing to the latest at Wusih—generally opened within a two-week period, between May 18 and June 4. Because the peasants were generally forced to sell within this time period, before the chrysalis emerged from the fresh cocoons, they had to sell at the prevailing price.[97] Although, in theory, the peasants could choose to reel their own silk if the cocoon price dropped too low, this was not a real option in areas which were new to sericulture, and where the scale of household production of cocoons was too large to make domestic reeling feasible.[98] Because competition for cocoons was keen in good seasons, and because the quantity sold by each peasant was usually small—from five to twenty catties—the result was hasty and careless buying, which contributed to the poor quality of cocoons.[99]

The brevity of the cocoon season created an annual drain on Shanghai's financial resources. Since purchases from peasants were usually made on a cash basis, during the season large quantities of silver were transported to the rural areas, usually about seventy-five million dollars (roughly fifty million taels) for the spring crop in Chekiang, Kiangsu, and Anhwei.[100] About 30 percent of the necessary cash came from Hangchow, Chia-hsing, Wusih, Soochow, and other places, while the rest came from Shanghai.[101] Both the steam filatures and the cocoon brokers borrowed from the *ch'ien-chuang* in order to purchase their cocoons.[102] Usually the borrower would put up 20–30 percent of the principal as a guarantee, and the cocoons as collateral. If the broker or filature borrowed 100,000 yuan, for example, he or it would receive only 70,000 yuan in cash to buy cocoons, the other 30,000 having already been deducted as a deposit.[103] The interest rate on such loans was between 6 and 12 percent.[104] Foreign capital was heavily involved in the financing of cocoon purchases. According to one estimate, of the fourteen to sixteen million taels worth of cocoons purchased each year for the Shanghai filatures, six million was put up by the *ch'ien-chuang,* two million by Chinese and foreign banks, and eight and a half million by foreign exporters. The procedure followed by the latter was similar to that of the *ch'ien-chuang,* and usually about 7.5 taels was charged as monthly interest on each 1,000 taels worth of cocoons.[105] Since foreign banks and firms often made short-term or chop loans to *ch'ien-chuang,* foreign capital was involved in the cocoon trade indirectly as well as directly.[106]

After the contract was made with the cocoon hong, the *ch'ien-chuang* would send a man to the district to pay the money and supervise the shipment of cocoons to Shanghai, where they became the property of the *ch'ien-chuang* and were stored in a warehouse designated by it.[107] If the price of cocoons fell later in the season, the deposit would be applied against the difference. It was possible to borrow from a modern bank at lower interest rates, but these loans were more difficult to negotiate. The cocoons stored in the warehouse were used by the *ch'ien-chuang* as a form of credit. Cocoon receipts were considered negotiable paper and were readily accepted by Chinese banks. They charged high interest rates, however, on loans negotiated with this paper. If the bank had financed the cocoons in the first place, the receipts were deposited at the bank and, whenever the purchaser wished to use any of the cocoons in storage, he had to pay the bank first.[108]

Except for the dry cocoons which were shipped directly to the filatures at Shanghai, the bulk of the cocoon supply was stored in the cocoon warehouses of which there were perhaps twenty in Shanghai, each having an average capacity of 5,000 piculs of dried cocoons, making a total capacity of about 100,000 piculs. This represented about 50–60 percent of the annual supply used by Shanghai filatures. The rest were stored either at the premises of the filature or in private warehouses.[109]

Since cocoon hongs operated only seasonally, and indeed in some years might not operate at all, their exact number at any given time is hard to reconstruct. According to one Chinese estimate, in 1917 there were a total of 682 hongs in the Kiangnan area: 456 in Kiangsu, 212 in Chekiang, and 14 in Anhwei.[110] According to Japanese sources, in the middle or late 1920s there were 1,228 cocoon hongs in Kiangnan: 782 in Kiangsu, 330 in Chekiang, and 116 in Anhwei.[111] During this period a substantial number of cocoon hongs were established for the first time in Hu-chou, Chia-hsing, Hangchow, and Soochow. These figures reflect the distinction between old and new sericultural areas, as discussed in Chapter 4 and shown in Table 19. The Steam Filature and Cocoon Guild of Kiangsu and Chekiang, formed around 1910 and expanded in 1913 to include Anhwei, tried to promote the interests of the steam filature industry by encouraging the expansion of sericulture into new areas. Shen Lien-fong, the head of the guild, was active in this effort, which received considerable support from foreign trading firms at Shanghai. He claimed that the traditional weaving interests had deliberately prevented the establishment of cocoon hongs in the old sericultural districts.[112]

Although there is no evidence of an organized conspiracy, the attempt of traditional interests to obstruct the development of cocoon hongs was reflected in provincial legislation limiting the number of hongs. In Kiangsu, regulations were drawn up around 1915 declaring that all those hsien with more than five cocoon hongs should not establish any other hongs for a period of five years. After this period, the building of new hongs was to be carefully regulated. In those hsien which did not yet have five hongs, new hongs could be built, up to the limit of five, except in Nanking and Soochow. In Chekiang, at about the same time, legislation was passed stating that within a radius of 20 *li* around centers already having cocoon hongs, no new hongs should be built. Where hongs were built, they should have no more than ten drying rooms.[113]

These provincial regulations, however, could not be completely enforced because the demand for cocoons was so strong, and in 1921, the National Law for Regulating Sericulture, Mulberry, Silk, and Weaving Industries in the Two Provinces of Chekiang and Kiangsu was promulgated by the Ministry of Agriculture and Commerce forbidding the building of cocoon hongs in the nine old silk districts—Hangchow, Chia-hsing, Shao-hsing, Hu-chou, Chiang-ning, Soochow, Ch'ang-chou, Chinkiang, and Sungkiang. The policy was designed to encourage the spread of sericulture to new areas. In those areas where hongs were permitted, their number would be restricted. In some areas like Sheng hsien in Shao-hsing, and in Wusih, public ovens were to be encouraged.[114] Hongs established before 1905 were allowed to continue operating, but no new ovens were to be added. In 1926, the Chekiang provincial assembly passed new regulations concerning cocoon hongs. Any steam filature of more than 100 basins established after that date would be allowed to own and operate 30 double drying ovens, or 45 single ones, in any three localities in the province. No tax would be levied on cocoons from Chekiang reeled in Chekiang.[115] Finally in 1928, after the establishment of the Nanking government, new regulations concerning cocoon hongs were promulgated, which, in theory, removed all restrictions. Payment of a 200-yuan fee would obtain permission to open a hong. As a result there was an immediate spurt in the number of hongs.[116]

Many contemporary observers felt that heavy taxation of cocoons was a key deterrent to the full development of steam filatures. Cocoon taxes, as distinct from silk taxes in general, were first imposed in 1883. In Chekiang, one picul of fresh cocoons was taxed at 4 yuan, and one picul of dry cocoons at 12 yuan. Of the latter, 9.4 yuan represented the regular tax, 2 yuan was for Shanghai tax, and 0.6 yuan was for transport. After the republic was established, both taxes were reduced slightly, and in 1916 they were further reduced. At that time the tax on fresh cocoons was 3 yuan per picul, and that on dried cocoons was 9 yuan per picul, plus 1 yuan surtax.[117] In 1931, the entire tax structure in Chekiang was revised, and cocoons, like other products, were assessed a commodities tax, the *ying-yeh shui*, of 0.2 percent.[118] This basic cocoon tax did not include numerous surcharges. According to one Japanese survey conducted in the mid-1920s, in Kiangsu province, the total tax on one picul of dry cocoons amounted to 12.80 yuan. This included 8.00 yuan regular tax, 0.50 surcharge, 0.30 guild tax, 0.10 reform association dues, 2.30 transit half-tax, and 1.60 special surtax. In Chekiang province, the total assessment amounted to 13.60 yuan, which included 9.00 yuan regular tax, 0.90 surtax, 0.30 guild dues, 0.10 reform association dues, 1.00 Whangpoo conservancy tax (the Shanghai tax), and 2.30 transit half-tax.[119]

Taxes on cocoons thus represented a substantial portion of the total cost of processing cocoons for filature reeling. According to the Japanese survey, the average cost of processing one picul of dry cocoons in Wusih was about 30 yuan, while the average cost at Shao-hsing was about 34 or 35 yuan. Taxes of 12.80 and 13.60 respectively thus represented about one-third of the total cost.[120] However, in terms of the final price of dry cocoons—which ranged from 170 to 372 yuan per picul—taxes represented only about 3–8 percent.[121] In terms of the total cost of producing one picul of raw silk for export, taxes amounted to 67.13 taels. This included 55.30 taels for cocoon tax, 11.14 for export duty, 0.36 for wharfage fees, and 0.33 for the Whangpoo conservancy tax.[122] According to Japanese observers this was shockingly high compared to the Japanese taxation of silk, but when viewed in light of the prevailing prices for filature silk around the mid-1920s, as shown in Table 14, the total tax of about 55 taels represented only about 5 percent—not as outrageous as the tax critics claimed. Under these circumstances it is difficult to see that the tax on cocoons posed a real obstacle to expansion or modernization.

The cocoon tax, however, was a major source of provincial revenue. In 1904 the Chekiang revenue from this source was only 197,154 yuan, but by 1925, it was 792,500 yuan, and by 1929, 1,345,644 yuan. By 1930, it was 1,765,158 yuan.[123] In Kiangsu, the total revenue in 1929 from the cocoon tax amounted to 1,286,347 yuan.[124] In short, the provincial governments of Chekiang and Kiangsu had a substantial interest in imposing a rather severe tax on the cocoon trade. It was in fact the single largest source of revenue for them.[125] Not only did the provincial government fail to provide any tax incentives to those who might want to start filatures, the central government, weak as it was in the 1910s and 1920s, took no steps to protect the silk reeling industry by preventing the export of cocoons. By the 1920s, the annual export of dry cocoons averaged 30,000–40,000 piculs a year, completely unrestrained by any protective tariff. This not only represented a significant loss to the Chinese industry, it also raised the price of cocoons. Almost all the cocoons were shipped to Japan, China's major rival in the silk trade.[126]

The cocoon hongs profoundly altered the position of the peasant household by separating sericulture from silk reeling. With their establishment in rural areas, the price of cocoons tended to rise rather sharply and, in times of prosperity, the peasant household could generally do better by selling their cocoons than reeling them at home.[127] In the old silk areas, the

peasants had the option of keeping their cocoons to reel themselves if the price dropped too low.[128] In new areas, however,
where the skills of silk reeling were not known, the peasants
were forced to accept even the lowest prices for their cocoons.
Moreover, since even a day's delay would decrease the weight
of the cocoons, they did not have any control over the timing
of their sales. On the other hand, when cocoon prices were very
high, filatures had a difficult time in realizing a good profit, and
some were forced to close down.[129]

Not only was the quantity of cocoons each season uncertain,
but it was agreed by foreigners and Chinese alike that their
quality suffered a marked deterioration with the establishment
of filatures. Since the peasant households no longer had to reel
the cocoons themselves, they were concerned only with the
quantity and weight of their cocoons, and their methods of
sericulture became more careless.[130] "The small peasant farmer
saw no motive in keeping up quality now that he sold his cocoons for ultimate delivery to steam filatures, or direct shipment abroad, in place of reeling them himself. The merchants to
whom he sold them were mere brokers, not particularly interested in quality provided they could resell at a profit what they
had bought."[131] As methods became more lax, the incidence of
disease in silkworm eggs spread, exacerbating an already serious
problem. Foreign merchants bemoaned "the suicidal tendency
of growers to pay more attention to quantity than quality."[132]

At Canton the situation was much the same despite the somewhat longer experience there with steam filatures. Methods of
breeding silkworms were careless, and disease was common-
place.[133] Agronomists from the Lingnan College Agricultural
School remarked, "The most surprising thing about Kwangtung
silk is that such good silk can be produced from such defective
cocoons."[134] The commerce in cocoons was handled in a somewhat different manner from Shanghai. Instead of cocoon hongs,
peasants sold their cocoons to cocoon markets (*chien-shih*) or
to cocoon warehouses (*chien-chan*). The former were owned by
private individuals or small companies. The market rented stalls

to brokers known as *shui-t'ou* (lit., "water-head"), who bought silk either on their own behalf as speculators or for particular filatures. There were an estimated 380 such merchants in the Canton area. The market earned a commission for each transaction of 3 percent from the seller and 1.5 percent from the buyer. The second way was to sell cocoons to warehouses which were owned by one or more filatures. Because the method of examination and sales was considered to be more scientific and fair, many peasants preferred to sell cocoons in this way. The warehouses sometimes extended loans to peasants, who were then presumably tied into the operation of the filatures.[135] The filatures were then responsible for the drying and processing of the cocoons. Since the Canton filatures were scattered about the countryside, their business with the foreign silk exporters was handled through commission houses, of which twenty or more were located at Shameen. They were organized as small companies, with an average capitalization of 50,000 yuan. The commission houses arranged for the storage and sale of the silk at Canton before export.[136]

It is actually rather surprising that Canton's silk should have been of such poor quality and that its silk filature industry should have shared so many of the problems which plagued the Shanghai reeling industry, for Canton possessed many advantages unknown to Shanghai. First, Canton had the important advantage of virtually year-round sericultural production. This should have eliminated many of the problems caused by the seasonal nature of sericulture in the Kiangnan area. Since most silk-raising peasant households in Shun-te and Nan-hai raised silkworms as their main occupation, the commercial and natural risks characteristic of the silk trade should have been minimized. Second, the Kwangtung filatures were located in the countryside, near the sericultural households. Presumably this proximity should have permitted the filatures better control of their sources of supply and perhaps to achieve some degree of vertical integration of the industry. The cocoon warehouses did help to establish a more direct relationship between some filatures and some peasant households, but the indirect method of purchase through *shui-t'ou* and the cocoon markets was the prevalent form of business. In Kiangnan, the concentration of steam filatures at Shanghai separated them from their rural source of supply and created costly bottlenecks. Because the filatures depended on middle-men and the cocoon hongs, they had very little control over the quality of cocoons they received. This is one reason why Wusih later became such an advantageous place to locate the filature industry. Third, the Canton filature industry did not have to contend with strongly entrenched traditional interests. The weaving industry in the delta was already sensitive to foreign markets and, although some competition between the two sectors was involved, it was in no position to resist effec-tively. In Kiangnan, by contrast, traditional interests and traditional practices did exercise a restraining influence on the development of the raw silk export trade—although by no means one which could not be overcome.[137]

The fact that none of these advantages made the Canton silk industry any stronger than the Shanghai industry suggests that the disadvantage that they shared may have been a more im-portant factor in the relative backwardness of the steam filature industry in China. This disadvantage was the scarcity of capital. This probably was not due to any absolute scarcity of available capital in China, although economists and economic historians disagree sharply among themselves on this point.[138] Instead it reflected the fact that the wealthy preferred to invest their money in real estate or in short-term, high-yield investments, such as pawnshops, than in development of long-term, high-risk modern enterprise. Among modern enterprises, silk reeling could be one of the most profitable but it was also one of the riskiest. More readily available capital might have made feasible vertical-ly integrated enterprises which could control some or all stages of the production process. But given the risks involved, it was easier for merchants to rely on those short-term middleman or brokerage transactions, which were relatively safe and were undertaken along familiar patterns. The rental of steam filature plants, which separated ownership from management, and the institution of cocoon hongs, were all modern variations on traditional commercial themes.

Does this then suggest that no real "modernization" had taken place? The fact that silk reeling had become organized in factories equipped with modern machinery and steam power was only technological modernization. Without measures that would help reduce the element of risk, such as modern banking and credit facilities, or a government-sponsored program to regulate silkworm egg production to reduce disease, technological change was insufficient. As George Allen and Audrey Donnithorne have remarked, "The silk trade, for its successful organisation needed a high degree of centralization at some points and a wide dispersion at others. The production of cocoons is necessarily a manual process and is highly suited to small-scale peasant agriculture. If, however, large quantities of uniform silk are to be provided... then there must be centralized supervision over certain processes."[139] In China the need for such centralized supervision in the form of government intervention was recognized by many people, but the efforts to provide it came too late to be effective.

REFORM MEASURES

In the reform of the silk industry, the process which was in greatest need of centralized supervision was the manufacture and distribution of silkworm eggs. The key to a filature's success was its supply of cocoons, and the key to the quality of cocoons was disease-free eggs. Disease was the main cause of the unreliable quality of Chinese cocoons and, although there had existed for decades the means of controlling it, the application of this technology required a degree of leadership and social persuasion which seemed to be lacking. Custom, not cost, was the main obstacle to spreading the use of "improved" eggs. The cost of egg-sheets was the smallest part of the total cost of sericulture, but according to the Shanghai International Testing House, "The ordinary farmer will not even take free of charge the egg-sheets of which he is suspicious, or which he has tried before and found to be poor."[140]

In the late Ch'ing, government efforts at reform were confined to the encouragement of the spread of sericulture by local officials and gentry and to the occasional modernizing efforts of high officials. These were isolated and uncoordinated attempts, largely directed at local welfare. Only toward the very end of the dynasty was there any public discussion of a national policy for promoting sericulture and silk exports for the national good. At Shanghai in 1896, a reform organization called the Nung-hsueh hui, or the Society for the Study of Agriculture, was formed under the leadership of Lo Chen-yü and others for the purpose of promoting technological improvements in agriculture.[141] The *Nung-hsueh pao,* which was published by this organization from 1897 to 1905, included such well-known reformers as Liang Ch'i-ch'ao, T'an Ssu-t'ung, and Chang Chien among its contributors. It devoted space to the translation of foreign works, including Japanese, on agricultural innovations. In the field of sericulture, it recognized the need for rapid improvement in order to reach the technological level of Japan, Italy, and France and to recapture the world market.[142]

During the Hundred Days of Reform in 1898, some attention was given to the problem of Chinese silk exports. Various proposals were put forth, including the establishment of a national institute of sericulture, the sending of students abroad to France and Japan to study the latest techniques, adoption of the Pasteur method of egg inspection, and free distribution of disease-free eggs.[143] Tuan Fang, who was appointed the chief minister for the National Bureau of Agriculture, Industry, and Commerce in July 1898, wrote a rather detailed analysis of the silk problem saying that the silk business suffered from insufficient capital. If business was good, the merchants could prosper, but when business was poor, the merchants went broke. Chinese silk merchants, even the more prosperous ones, were at a disadvantage when dealing with the foreign exporters. The only way of protecting the silk merchants, he said, was to establish a government bank and a joint government-merchant company which would protect the interests of the silk business.[144] Unfortunately, however, Tuan Fang's proposals and those of the other reformers were swept away with the Hundred Days of Reform.

The reform efforts of the early twentieth century were thus left by default to certain individuals and educational institutions. One of the earliest and most successful private reform efforts was the establishment of a sericultural school at Hangchow by a local magistrate, Lin Ti-ch'en, with the encouragement of F. Kleinwachter, the customs commissioner.[145] There were Japanese instructors,[146] but the real force behind the school came from the students, returned from study abroad, who reorganized it around 1909.[147] It was later renamed the Provincial Sericultural School of Chekiang and received a 200,000 yuan annual appropriation from the provincial government. It had 150 students at the middle-school level. By the 1920s, this school had become the major center for the production of scientifically prepared eggs. Each year 17,800 to 23,800 sheets were produced and marketed mostly outside of Kiangnan —in Shantung, Honan, Hupei, and Anhwei. These provinces also sent many students to the school.[148] Because of this school, Hangchow became the leading producer of "improved" eggs in the province.[149]

Another successful venture was the Hushukwan Girls' Sericultural School (Hu-shu kuan nü-tzu hsueh-hsiao) near Soochow, established in 1913. This was the largest and best known of such institutions in China, and in the 1920s it received a 35,000 yuan annual subsidy from the Kiangsu provincial government. Free tuition and board were given to students from within the province, and a nominal fee was charged to those from outside. The students and instructors of the school manufactured and distributed egg-sheets in an active program of extension work, in which this school was considered a pioneer.

Many of its graduates became important leaders in the silk industry.[150]

The University of Nanking established a Department of Sericulture in 1918, which conducted research on silkworm egg disease and maintained an experimental station with mulberry fields. This school, like the university itself, had close ties with the United States, and the U.S. Silk Association contributed money toward its operation.[151] In later years it ran in Kiangsu four highly successful demonstration centers which distributed egg-sheets to peasants.[152] There were a number of other schools and reform associations established in Chekiang and Kiangsu which were not so successful. Many of them suffered from poor equipment and incompetent management.[153] Some functioned in name only. Returned students from Japan sometimes tried to use their training to set up sericultural schools, but often they encountered anti-Japanese sentiment and also competition from students returned from the United States and Europe.[154] In south China, Lingnan Agricultural College was active in sericultural reform.[155]

Several projects promoted by the provincial governments in the early republic ended in failure.[156] In 1925 the Chekiang provincial government established five model reform districts with several stations to distribute "improved" eggs, but according to one official account, this project became "influenced by politics" and was disbanded.[157] After the Nationalist Government became established at Nanking in 1927 and began to support sericultural reform, the provincial government of Chekiang began to take more serious steps toward reform than it had before. A provincial sericultural reform station was established and, in 1928 and 1929, a number of experimental factories and model stations were set up, particularly in the old sericultural districts.[158] These experimental stations were reported to be very successful, especially in the 1929 season. Private egg manufacturing and distribution were encouraged to supplement the supply of the government stations. In 1931, however, after the effects of the depression were felt, the provincial government cut its sericultural reform budget from 330,000 yuan to 50,000.[159]

These various efforts to introduce "improved" eggs began to yield results in the 1930s. Kiangsu province had 125 privately owned egg-production centers, while Chekiang had 21.[160] In most of the silk districts, peasants came to accept "improved" eggs in their sericultural work. In Chia-hsing, for example, families which used "improved" eggs rose from only 5 percent in 1928, to 25 in 1931 and 1932, and to 50 percent by the spring of 1935. This demonstrated that concerted extension work could overcome the prejudices of the peasants.[161] In Wu-hsing, the percentage of families using only "native" breeds declined from 76 in 1931 to 61 in 1934. The percentage of families using both "native" and "improved" eggs rose from 19 percent in 1931 to 32 in 1934. The percentage of families using only "improved" breeds rose from 5 percent to 8 in the same period.[162]

There is evidence, however, that the model districts and other provincial reform efforts were subject to much political abuse. At Wusih, for example, it was widely felt that the influential Hsueh family, which controlled an empire of filatures and co-coon hongs, used the model district program to complete their domination of the local silk market. The Hsueh family was descended from Hsueh Fu-ch'eng, who had served as Minister to England and other European countries in the 1890s.[163] Hsueh Fu-ch'eng's son, Hsueh Nan-mo, became a prominent dealer in cocoons and investor in silk filatures. In 1909 he helped found the Chin-chi filature and became a leader of the Wusih clique of businessmen at Shanghai. The family had also acquired interests in the Yung-t'ai filature, which was one of Shanghai's largest and most successful. In 1925, this factory was moved to Wusih.[164] After the depression, half of Wusih's filatures were forced to close down, but Hsueh Nan-mo's son, Hsueh Shou-hsuan, was able to effect a revival of the family's fortunes by first renting a number of filatures and then trying to control the cocoon crop.[165] Their critics said that Hsueh

and his colleagues had the power to control the price of cocoons at Shanghai and to force the raw silk price up five times
or more. According to a Wusih newspaper, Hsueh made two
million yuan or more this way.[166] According to one source, only those on friendly terms with the family were allowed to
operate cocoon hongs, and the peasants could sell their cocoons
only to these hongs. The cooperative movement fell under this
family's control as well, and the provincial government named
Wusih a model sericultural district. The reform movement was
a pretext by which the Hsueh family expanded its control over
the Wusih silk industry. By 1937, they controlled about two-
thirds of Wusih's cocoon hongs and filatures.[167]

The Chekiang provincial government decided to encourage the
growth of the cooperative movement after the dual disasters
of the depression and the Japanese attack on Shanghai.[168] By
1933–1934, Chekiang had 205 spring and 226 autumn silkworm
cooperatives. Each cooperative society had sixteen members,
and it was estimated that about 8 percent of the sericultural
families in the province belonged to one.[169] In Kiangsu there
were 101 such cooperatives.[170] The most famous silk cooperative was that at Kai-hsien kung, south of the T'ai-hu near Chen-
tse, which was described by Fei Hsiao-t'ung in his *Peasant Life
in China*. This cooperative involved not only the joint raising
of silkworms, at least in the early stages, but also the establishment of a cooperative reeling factory, which was begun as an
experiment in 1929. By 1935, its silk was given the highest
rating by the testing house. The cooperative failed, however, to
provide enough financial incentive for its members to supply it
with a sufficient amount of cocoons and had to depend on the
free market for the rest of its supply. The entire venture was
financed by the Provincial Peasant Bank. This, combined with
the fact that it was the policy of the Nanking government to
promote the cooperative movement as part of its rural reconstruction program, suggests the importance of political support
for any type of fundamental reform in the silk industry.[171]

International support and initiative also played a significant role in sericultural reform. In 1917 and 1918, the Foreign Silk Association of Shanghai and the Steam Filature and Cocoon Guild of Kiangsu, Chekiang, and Anhwei, joined together to establish the International Committee for the Improvement of Sericulture in China. The French, British, Japanese, and American Chambers of Commerce also supported this effort, and each contributed 5,000 taels to its financial upkeep. The Chinese government contributed 4,000 taels a month from the Customs revenue. In 1922 this sum was raised to 8,000 taels a month.[172] The main activity of this committee in its early years was the distribution of scientifically prepared eggs to peasants. It set up production stations at Shanghai, Chia-hsing, and a number of key places, but the greatest portion of the egg-sheets that it distributed was purchased from France and Italy. Later the committee made strenuous efforts to promote the Chinese production of disease-free eggs.[173]

American silk dealers were particularly active in advocating sericultural reform in China. In 1920 when an American delegation of silk businessmen visited China, it proposed that the United States and China jointly organize a silk testing station in Shanghai. In the following year, a Chinese delegation visited the United States to attend the International Silk and Satin Exposition, and there the American and Chinese delegations decided to raise 60,000 taels each to establish this testing house.[174] The Shanghai International Testing House was established in 1922 to provide testing services for Chinese filatures and foreign exporters.[175] This public service was welcomed by firms at Shanghai, which previously had to rely on their own inspection departments, and used visual, rather than mechanical, tests to grade their silk.[176] The Testing House also supported the reform efforts at Nanking University and Lingnan College in Canton.[177] After the Nanking government was established, it gradually gained control over the Committee for the Improvement of Sericulture. And in 1930, the Ministry of Industry assumed control of the International Testing House as well, which was renamed the Chinese Sericultural Reform Association.[178]

In addition to these official efforts, there were some private efforts at reforming the methods of financing and marketing in the silk trade. In the late 1920s and early 1930s, there were about five attempts to set up Chinese silk exporting companies, all of them short-lived.[179] In 1925, a number of Hu-chou silk merchants, under the leadership of Chou Hsiang-ling, established the Hu-chou Model Filature. Although this venture survived only until 1932, it had an important impact because it sold its silk directly to foreign buyers. In 1928, the Hu-chou Model Filature joined with the Nan-hsun Silk Filature, which was also founded by Chou Hsiang-ling, and with the Hangchow Filature and Hou-sheng Filature at Chia-hsing, to form a consortium to market silk directly in the United States and Europe. This group was organized as a division of the Tonying Trading Company, which already had branch offices in New York and Paris for the purpose of dealing in Chinese antiques. The success that this venture had in by-passing the usual foreign exporting firms and dealing directly with mills in the United States and Europe encouraged several important Shanghai and Wusih filatures—Jui Lun, Yung T'ai, Chien-sheng, and Chen I—to join the Tonying Company.[180]

These various reform efforts failed to effect any fundamental improvement in the silk industry, in part, because they were initiated too late. But essentially they failed because they represented a case of partial remedies, just as the steam filatures represented a case of partial modernization. The efforts by local officials to promote sericulture were based on a narrow view of local self-sufficiency. The efforts to promote the use of disease-free eggs were of great importance, but they attacked the technological problem without touching the commercial structure within which technological change had to be carried out. The International Testing House tried to impose some type of standardization, but also failed to deal with commercial practices. All of these reform efforts failed to focus attention on the underlying economic structure within which the silk trade was conducted. The cooperative movement was the single exception to this because it gave the peasants some leverage when dealing with the cocoon and silk markets, as well as some means of technological improvement. Unfortunately, it too was too little too late.

Conclusion

Silk was a unique and valuable product which played an important role in the Chinese economy of the modern period, as it had in earlier times. With the expansion of the European market in the mid-nineteenth century, and the opening of the American market in the late nineteenth, new opportunities for the export of Chinese silk presented themselves. Exports of raw silk and silk fabric to these foreign markets rose in value from 18 million HK taels in 1868 to a high of 176 million in 1926. Although silk's position in China's overall trade declined, the earnings from its sale formed a vital part of China's trade picture. During this period, the import of cotton yarn and cloth—widely thought to have threatened the livelihood of Chinese peasant families—increased from less than 2 million and 22 million HK taels in value respectively to 28 and 205 million taels. In effect the value of silk exports just about offset the value of cotton cloth imports. Put another way, if China had not exported silk, its balance of trade—unfavorable in every year between 1865 and 1937 with the exception of 1872–1876—would have been far more seriously unbalanced.[1]

The large-scale expansion of silk output, essentially new to many areas, such as Kiangsu, Kwangtung, and Shantung, brought a new source of income to peasant households in these regions. In the old districts in northern Chekiang, it brought revived prosperity as more households undertook silk production and the value of their output rose. Sericulture was ideally suited to China's peasant economy. It required relatively little capital. Its technology was not complicated, and it did not require complex machinery. It could be profitably undertaken on either a small or large scale. More important, sericulture took full advantage of China's most abundant resource, its man and woman power. The highly labor-intensive requirements of sericulture were perfectly suited to the rural household situation, particularly where sericulture was seasonal. In the industrial sector, mechanized reeling was also potentially attractive because its technology too was relatively simple and its capital requirements relatively modest.

However substantial the contribution of the silk export trade to China's modern economy, the question which has been most frequently asked of it is why it could not have been still more substantial. If China had been able to capture the American market which the Japanese eventually dominated, China could have earned an even larger income, it would not have had a balance-of-trade problem, and Chinese peasants and merchants would have derived even greater benefits. For all of this to have happened, however, the silk industry would have had to modernize more smoothly, which in turn would have been possible only if China had not experienced a modernization problem to begin with. This type of counterfactual reasoning may be futile, but the experience of the silk industry does serve to reveal the strengths and weaknesses of the Chinese economy.

Whatever advantages silk manufacture offered to China's

economy, it also had some serious shortcomings. The sensitivity of silkworms to natural conditions and the increasing vulnerability of Chinese silkworms to disease introduced a very large element of risk into the sericultural process. This risk was compounded by the instability of both the domestic and international markets for silk. These combined risks meant that peasant households, except in the Canton delta, limited their output of silk and that merchants and industrialists limited their investment in the silk trade and in filatures. The traditional system of household production and the decentralized networks of manufacture and commerce, which were already well-suited to risk minimization, were adapted for use in new contexts, such as factory management, with relative ease. Thus, the experience of the silk trade shows that the Chinese economy, in both its agrarian and commercial sectors, was quite capable of response to economic opportunity but was equally capable of protecting itself against unnecessary risk. Meanwhile, the domestic market for silks remained steady after dropping from its early Ch'ing heights, and this contributed to the muting of the response to the foreign market, further reducing the incentives for structural change.

Strong government leadership could have overcome many or most of these disadvantages. Disease, the most fundamental technical problem, could have been eliminated only through direct government regulation of egg production and distribution. The government could also have helped peasants by extending easy credit and by encouraging, as it did in the 1930s, the formation of rural producers' cooperatives, which would have reduced the vulnerability of individual households and given cohesiveness to otherwise atomized units of production. In industry, the government could have fostered the expansion of silk reeling filatures by providing institutional support for growth. The filatures, along with modern enterprises in general, suffered from the absence of a modern banking system, just as the export trade in general suffered from the Chinese failure to seize control of shipping from Western companies.

Since Japan was China's chief rival in the world silk market, it might be instructive to look briefly at the development of its silk industry.[2] Its spectacular growth has usually been attributed to an aggressive government policy of developing Japan's export capability. To promote the growth of silk exports, the Japanese government not only set up model filatures and inspection stations, and regulated the breeding and distribution of silkworm eggs, but it also encouraged the extension of credit through a modern banking system. With the establishment of the Yokohama Specie Bank in 1880 and the Bank of Japan in 1882, commercial banks were able to assist the Yokohama silk wholesalers in making loans to silk reelers.[3] By the first decade of the twentieth century, bank loans represented about 75 percent of all silk financing and wholesalers' loans about 25 percent.[4] The silk trade in fact contributed to the development of Japanese banking, just as the banking system was a critical factor in drawing capital into the silk business.[5] Behind the banking system stood the Japanese government, which was directly responsible for a substantial portion of the financing of the silk trade.[6]

Although government leadership has usually been cited as the key factor in the development of the Japanese silk industry, the general economic and historical context was equally important. Foreign trade in general played a larger role in Japan's economy than in China's and, in contrast to its Chinese counterpart, the Japanese silk industry was far more directed toward the foreign market than the domestic. From 1883 to 1937, between 58 and 78 percent of Japan's output of raw silk was exported, usually well over 65 percent in any given year.[7] Japanese silk production was also more geographically extensive than the Chinese, with some silk being produced by almost all prefectures in the twentieth century.[8] At the height of its development in the 1920s, 40 percent or more of all Japanese farm families were engaged in sericulture, whereas in China such a degree of involvement was achieved only in a few specialized localities.[9]

In Japan, moreover, the opening of trade with the West was preceded by a period of internally stimulated growth and innovation in sericulture, whereas in China it coincided with a period of technological stagnation and decline of government demand. During the seventeenth century, the Tokugawa Bakufu, having tried unsuccessfully to curb the rising imports of silks from China, encouraged the promotion of sericulture by various han governments to meet the demands of the domestic market. Financial incentives, the distribution of mulberry seeds and silkworm eggs, and the publication of sericultural manuals were among the methods employed.[10] These efforts produced striking results. By the late Tokugawa period, sericulture had been introduced to most parts of Japan, except the north. By the mid-nineteenth century, the Japanese had not only caught up with the Chinese in silk technology, but had begun to experiment with some techniques yet unknown in China.[11] The *zaguri* or "sitting" type of reeling machine, already widespread in China, was introduced to Gumma and Fukushima, replacing the hand-operated machine.[12] Other technological innovations included the cold storage of eggs and the development of summer and fall crops of worms.[13] Although Chinese techniques were still more advanced on the whole, there was no comparable innovative activity in China in this period. Moreover, in Japan these technological changes were accompanied by changes in the organization of production. For example, as early as the mid-eighteenth century in Gumma prefecture, workshops were set up outside the household, thus separating reeling from sericulture.[14] In China such separation of functions did not occur until the advent of steam filatures.

In short, a considerable momentum had been generated in Japan even before the opening of Western trade. In the 1850s and 1860s, several new centers of sericulture emerged. Fukushima and Gumma prefectures, traditionally the sources of the best raw silk, were joined by Nagano, Gifu, and Yamanashi. The manufacture of egg-cards for export to Europe flourished in this period. Local entrepreneurs disseminated knowledge of sericultural techniques.[15] After the Meiji Restoration in 1868, a mixed pattern of traditional and modern production began to emerge.

In the 1870s, several pioneering ventures in steam filatures were launched, the most important of which was the government model factory, the Tomioka, built in Gumma prefecture in 1872. This project was directed by a French expert, accompanied by eighteen European assistants, and used the latest foreign machinery. Although ultimately a business failure, the Tomioka set an important example, and its staff were dispatched as instructors to other localities that wanted to learn modern techniques.[16]

The actual spread of steam filatures was, however, rather slow. Many enterprises classified as "factories" were in fact workshops having only ten to thirty workers. Moreover, in contrast with the larger-scale European-style filatures, they used water power, not steam.[17] The adoption of Western technology was a complex process, with various regions responding differently. Gumma and Fukushima, for example, accepted Western technology later and with greater difficulty than the new silk districts such as Nagano.[18] In Gumma especially, as in Kiangnan later, traditional technology was improved to produce *kairyō-zaguri ito*, domestic silk rereeled to produce a more even thread.[19] It was not until the turn of the century that modern machine-reeling assumed the dominant role in the production of Japanese silk. In 1900 only 53 percent of Japan's total output of raw silk was machine-reeled; by 1925 well over 75 percent was, including virtually all exported silk.[20] The modern filatures were larger in scale than the previous workshops, and they increasingly used steam power.[21] After the 1910s, two giant companies, the Katakura and the Gunze, dominated the machine-reeling industry, branching out from their respective bases in Nagano and the Kansai to establish factories elsewhere.[22] But despite the spread of mechanization and the rise of large firms, on the whole the organization of Japanese silk production retained its small-scale character. In 1926, for example, 36 percent of all filatures had fewer than 50 basins; 66 percent had fewer than 100 basins.[23]

Another characteristic of the Japanese silk industry was the continued strength of its rural sector, even after the industry was largely modernized.[24] Much of the capital for the early ventures in filatures came from rural entrepreneurs.[25] And from the early Meiji period, rural silk merchants formed cooperatives to weigh, pack, and ship silk to Yokohama.[26] By the Taishō period (1912-1925), peasant producers joined into village-wide unions in order to negotiate collectively with filatures, or to form cooperatives for reeling their own cocoons. With government support, the cocoon unions formed federations within each prefecture, and a nationwide federation appeared in the early Shōwa period (1926-).[27] Because filatures were relatively small, they were well distributed throughout the rural silk districts, closer to their source of cocoons. With the advent of large companies, it became increasingly common for filatures to distribute silkworm eggs to peasants—an even more effective way of insuring the quality of their cocoon supply.[28] Thus, rural cooperatives provided a type of horizontal integration of the silk industry, giving a measure of bargaining power to otherwise vulnerable rural producers, while large firms vertically organized the industry, involving themselves in almost all stages of production in order to insure themselves a better product.

The development of the Japanese silk industry and its preeminence in the international market by the early twentieth century, while dramatic in appearance, were based on factors that have not been well-publicized. It was not merely a story of strong government, big business, and rapid technological change, but also a story of significant early modern momentum, steady incremental growth on a small scale, and the continued strength of the rural sector and its integration with the industrialized sector.

Recently it has been argued that historians have been wrong in attributing the relative ease of Japan's transition to industrialization, and the relative difficulty of China's, to cultural and institutional differences. Rather it has been asserted that China and Japan had broadly similar societies and institutions, and the primary cause of their contrasting responses to the challenge of change was the fact that China was much more "incorporated" into the world economy, while Japan remained relatively free from "incorporation," another term for Western imperialism, in the nineteenth century and therefore had a short "breathing space" in which to take bold steps toward industrialization. The Chinese government tried to do the same, it is argued, but had been so politically and financially weakened by the Western powers that it could not take effective action such as subsidizing its new industries.[29]

While this is not the place to discuss the validity of this argument for modern Chinese and Japanese economic history as a whole, it is appropriate to determine its applicability to the silk export trade in particular. To what extent was foreign imperialism a factor impeding the development of the Chinese industry and encouraging the development of its Japanese counterpart? Although one might argue that the Chinese central government during the Republican period was too burdened by foreign debts to take a more active role in directing the development of sericulture, the performance of the Ch'ing government cannot be excused on these grounds. Throughout the late nineteenth century, government officials reacted along traditional lines to the opportunity for expanding the silk export trade and had not even begun to conceive of a protectionist or nationalist policy. Bureaucrats had traditionally played an active role in promoting sericulture, and the Ch'ing government had retained a faith in the idea of rural self-sufficiency and an essential distrust of foreign trade.

The silk filature industry was in fact the prime example among China's modern enterprises of development free of direct foreign investment. With one or two early exceptions at Canton and Shanghai, and the later exception of the Japanese in Shantung and Manchuria, the silk filature industry was developed by Chinese capital. The positive benefits which might have been expected from this, however, did not develop, as the economic uncertainty of this business encouraged the tendency to keep as much of one's assets liquid as possible, and "native" capital always seemed scarce. Moreover the indirect dependence of filatures on foreign banks and export companies for their working capital perhaps reinforced the speculative aspects of this business. The foreigners seemed also to be trying to minimize their risks by not investing in fixed assets. In that sense, and only in that sense, it might have been better for the stability of the business if more direct foreign investments had been made instead of short-term financing.

The filature industry was to a large extent a treaty port phenomenon. In the case of Shanghai it suffered from being separated from its rural source of supply. The advantages which Shanghai could offer, including easy access to Chinese and foreign capital, drew businesses there, and in so doing created an apparent dualism. In the Canton delta, however, where filatures were located in the countryside near their source of cocoons, and not in the treaty port of Canton, there was no appreciable superiority in the final product or in the management of the business. The same type of insecurity and instability that characterized the Shanghai business also existed in the Canton delta. From this we might conclude that, while the presence of the treaty port may have promoted a dualism at Shanghai, the dualism did not sufficiently explain the troubles of the industry.

The problem was not that foreign investment dominated the silk industry (it did not), nor that the treaty system forced a split between urban industry and rural hinterland (even though it sometimes did, it was not the critical problem), nor that the Ch'ing government was restrained by financial impoverishment from promoting the industry (there is not much evidence that it wanted to). The problem was that there were no steps taken to overcome risk and uncertainty either by collective action from below or by government action from above.

In short, the concept of imperialism or incorporation is not very useful for an analysis of the silk trade and cannot explain the difference between the Chinese and Japanese performances. Although the world economy or imperialist perspective has concentrated on trade in those commodities where a dependent relationship was established and which one might argue had unfavorable effects on the Chinese economy, such as opium or cotton, it has failed to take into account those commodities which might have, or actually did have a favorable impact.[30] The Chinese and Japanese experiences with silk both suggest that trade could have extremely beneficial consequences, both in concrete economic terms and in long-run developmental terms. The Chinese notion of self-sufficiency and the orientation to a substantial domestic market, however, muted the response to this opportunity for trade, whereas the greater Japanese dependence on trade, as well as the more aggressive Japanese attitude, sharpened the Japanese reaction to the world market. When the world depression occurred, and the silk market eventually collapsed, Japanese peasants perhaps suffered more keenly than their Chinese counterparts, but no Japanese historian to my knowledge has ever argued that Japan should therefore never have developed its silk trade in the first place.

It does not necessarily follow, however, that the contrasting experiences of the Chinese and Japanese silk industries in this past century should be attributed entirely to institutional or cultural differences. As the nineteenth century recedes into the past, the similarities between Chinese and Japanese institutions and practices may indeed come to appear more striking than their differences, and their common strengths more important than their weaknesses. Shorter-term historical and economic factors, however, will play a larger role in explaining short-term phenomena. It is only when we view the Chinese experience in juxtaposition with the history of the Japanese silk industry, and in the context of the international market during this relatively limited period of time, that the Chinese industry seems backward or unsuccessful. In this sense the silk export trade typified one of the major dilemmas of modern China's history—how to survive a highly competitive international struggle in which it was a reluctant participant.

Abbreviations
Notes
Bibliography
Glossaries
Index

Abbreviations

AR	*Annual Report.* Silk Association of America
ATRR	*Annual Trade Report and Returns.* Comp. China, Maritime Customs
CC	*chen chih* (gazetteer of a *chen*)
DR	*Decennial Reports.* China, Maritime Customs.
FC	*fu chih* (gazetteer of a prefecture)
HC	*hsien chih* (gazetteer of a hsien)
HTS	*Hu ts'an-shu*
KTSSCP	*Kuang ts'an-sang shuo chi-pu*
KYSTL	*Chung-kuo chin-tai kung-yeh shih tzu-liao*
KZ	*Shina keizai zensho*
NYSTL	*Chung-kuo chin-tai nung-yeh shih tzu-liao*
PNS	*Pu Nung-shu*
Silk	*Silk.* Comp. China, Imperial Maritime Customs
SK	*Shina sanshigyō kenkyū*
SKYSTL	*Chung-kuo chin-tai shou-kung-yeh shih tzu-liao*
ST	*Shina sanshigyō taikan*
Survey	*A Survey of the Silk Industry of Central China*
SYC: Che-chiang	*Chung-kuo shih-yeh chih: Che-chiang sheng*
SYC: Chiang-su	*Chung-kuo shih-yeh chih: Chiang-su sheng*
SZ	*Shina shōbetsu zenshi*
TSCMCS	*Ts'an-sang chien-ming chi-shuo*
TSYL	*Ts'an-shih yao-lueh*

Notes

Introduction

1. Kwang-chih Chang, *The Archaeology of Ancient China,* 3rd ed. (New Haven, 1977), p. 95. Cocoons have been found in neolithic sites in north China.

2. Ray Huang, *Taxation and Governmental Finance in Sixteenth Century Ming China* (Cambridge, England, 1974), p. 136.

3. According to some sources, this was particularly the case in Kiangnan, where the tax burden was especially heavy. See Hatano Yoshihiro, *Chūgoku kindai kōgyōshi no kenkyū* (Tokyo, 1961), p. 20.

4. Chang K'ai, *Ts'an-yeh shih-hua* (Peking, 1979), pp. 6–7.

5. Chao Ya-shu, "Sung-tai ts'an-ssu-yeh ti ti-li fen-pu," *Shih-yuan* 3:77–80, 85–87 (September 10, 1972).

6. Yoshinobu Shiba, *Commerce and Society in Sung China* (Ann Arbor, 1970), pp. 111–121, provides good descriptive material about silk commerce in the Sung.

7. Chang K'ai, pp. 21–22.

8. Wang T'ao, *Man-you sui-chi,* in *NYSTL,* I, 428.

9. Ying-shih Yü, *Trade and Expansion in Han China: A Study of the Structure of Sino-Barbarian Economic Relations* (Berkeley and Los Angeles, 1967), pp. 164–165.

10. Ibid., pp. 152–166, 176–177; G. F. Hudson, *Europe and China* (London, 1931), pp. 66–68, 77–82, 90–93, 120–123.

11. Hudson, pp. 155–160; R. S. Lopez, "China Silk in Europe in the Yuan Period," *Journal of the American Oriental Society* 72.2:73–74 (April-June 1952).

12. Ying-shih Yü, p. 47; Henry Serruys, "Sino-Mongol Relations during the Ming, II: The Tribute System and Diplomatic Missions (1400–1600)," *Mélanges chinois et bouddhiques* 14:213–229 (1967).

13. William W. Lockwood, *The Economic Development of Japan,* rev. ed. (Princeton, 1968), p. 94.

14. Kenzō Hemmi, "Primary Product Exports and Economic Development: The Case of Silk," in Kazushi Ohkawa, et al., eds., *Agriculture and Economic Growth: Japan's Experience* (Princeton, 1970), p. 308.

15. This view is forcefully argued in Rhoads Murphey, *The Treaty Ports and China's Modernization: What Went Wrong?* (Ann Arbor, 1970).

16. See Kang Chao, *The Development of Cotton Textile Production in China* (Cambridge, Mass., 1977), and Albert Feuerwerker, "Handicraft and Manufactured Cotton Textiles in China, 1871–1910," *Journal of Economic History* 30.2:338–378 (June 1970).

ONE   *The Technology of Silk*

1. Sung Ying-hsing, *T'ien-kung k'ai-wu: Chinese Technology in the Seventeenth Century*, trans. E-tu Zen Sun and Shiou-chuan Sun (University Park, 1966), pp. 37–39, says that this was common practice in Szechwan but rare in Chekiang, but *KTSSCP*, shang 11b–12a and *TSCMCS*, 13a both suggest that this was not uncommon in Kiangnan in the nineteenth century. The *che*, or "silkworm thorn" is the *Cudarica triscupidata*. I am grateful to Francesca Bray, Gonville and Caius College, Cambridge University, England, for this information.

2. Charles Walter Howard and Karl P. Buswell, *A Survey of the Silk Industry of South China* (Hong Kong, 1925), pp. 41–42.

3. Leo Duran, *Raw Silk: A Practical Handbook for the Buyer*, 2nd ed. (New York, 1921), pp. 13–14.

4. *TSCMCS*, 5a.

5. *KTSSCP*, preface 1a, 1:1a; *TSCMCS*, 5a; Chiang Yü-ch'ang, *Sang-ts'an shuo*, in *NYSTL*, I, 605.

6. D. K. Lieu, *The Silk Industry of China* (Shanghai, 1941), p. 3, found in his study of the Wu-hsing district of Chekiang that the soil was 94.5% loamy. Robert Fortune, *A Residence Among the Chinese: Inland, on the Coast, and at Sea* (London, 1857), p. 343, also commented on the loamy quality of Chekiang soil. Tadao Yokoyama, *Synthesized Science of Sericulture* (Bombay, 1962), pp. 12–14, discusses the importance of loamy soil.

7. I am grateful to Professor Rhoads Murphey, a specialist on Chinese geography at the University of Michigan, for explaining this point to me in a letter of April 13, 1973.

8. *Chia-hsing FC* (1879), 32:22a–b; *Nan-hsun CC* (1922), 30:19a.

9. *KTSSCP*, 2:48a.

10. Robert Fortune, *Two Visits to the Tea Countries of China* (London, 1853), I, 274; II, 14. See also his *A Residence Among the Chinese*, p. 343.

11. *ST*, p. 73.

12. *TSCMCS*, 11b–12a; *TSYL*, 9a.

13. *KTSSCP*, 1:4a.

14. *ST*, pp. 28, 73.

15. *TSYL*, 8a–b; and *Kuei-an HC* (1882), 11:5b.

16. *Silk*, pp. 46, 51. *ST*, p. 63.

17. Layering was the common practice during the sixth century according to the earliest extant agricultural manual, the *Chi-min yao-shu*, as Francesca Bray, who is preparing a translation of this work, has observed. Evidence of the widespread practice of transplanting saplings, and the risks incurred during the transportation, can be found in *HTS*, p. 11; *TSYL*, 5b–6a; *KTSSCP*, 1:7b, 12a–14a; and *Nan-hsun CC* (1922), 30:25a.

18. *PNS*, 1:13a.

19. *Kuei-an HC* (1882), 11:4b; *ST*, p. 62.

20. Ch'en Heng-li, *Pu Nung-shu yen-chiu* (Peking, 1958), pp. 193–196.

21. Robert Fortune, *Three Years' Wanderings in the Northern Provinces of China, Including a Visit to the Tea, Silk, and Cotton Countries* (London, 1847), p. 359.

22. *TSCMCS*, 8a; *HTS*, pp. 13–14; *TSYL*, 7b–8a; *KTSSCP*, 4b–7b; and *Silk*, pp. 47, 70.

23. *ST*, pp. 73–74; and Yueh Ssu-ping, *Chung-kuo ts'an-ssu* (Shanghai, 1935), p. 20.

24. Ch'en Heng-li, pp. 196–198; *PNS*, 1:11b–12a; *KTSSCP*, 1:8a–9a; *TSCMCS*, 14b–16a.

25. Ch'en Heng-li, p. 251; *TSCMCS*, 9a–10a; and *SYC: Che-chiang*, IV, 173–175.

26. The taller trees in northern Chekiang were planted about 140–150 trees per mou, while the shorter trees of Wusih were planted about 300 trees per mou. See *Survey*, pp. 27 and 49; and *ST*, pp. 73 and 82.

27. Ch'en Heng-li, p. 38. See also *KTSSCP*, 1:1a–b.

28. *Chia-hsing FC* (1879), 32:24a. Tadao Yokoyama, p. 250, concurs that the type of soil and the intensity of cultivation were the two important variables in mulberry yields. In Japan they could cause the yield of silk per acre to vary by a factor of six.

29. *Survey*, p. 6.

30. Howard and Buswell, p. 66; D. K. Lieu, p. xii.

31. Interview with Professor Marian Goldsmith, Department of

Developmental and Cell Biology, University of California at Irvine, on January 14, 1974, in Cambridge, Massachusetts.

32. *PNS*, p. 20; Ch'en Heng-li, p. 243.

33. *TSCMCS*, 39b; 38b–40a, gives a detailed description of how the summer crop should be treated.

34. *Shuang-lin HC* (1917), 14:11b. Also *Chia-hsing FC* (1879), 32:37b. Yueh Ssu-ping, pp. 27–28, shows the increasing importance of the summer crop in the twentieth century.

35. *Wu-ch'ing CC* (1936), 7:18b–19a; *SYC: Che-chiang*, IV, 166, 183.

36. *Survey*, p. 7; *TSCMCS*, 19a. In south China, the time from hatching of the eggs to spinning of the cocoons was only 16–17 days. See "Kwang-tung Silkworms," *Chinese Economic Journal* 5.2:723 (August 1929).

37. *Chia-hsing HC* (1906), 15:9b; *Kuei-an HC* (1882), 11:8a and 12:10b; and *HTS*, p. 44.

38. *KTSSCP*, 2:5b–6a; *TSCMCS*, 22b.

39. *KTSSCP*, 2:7b–15b.

40. Sung Ying-hsing, p. 38; *KTSSCP*, 2:9a, 16a; *TSYL*, 11b–12a.

41. Sung Ying-hsing, p. 41, and *TSCMCS*, 18b–19a.

42. Sung Ying-hsing, p. 42. Sometimes screens alone, or clumps of grass, were used instead of straw stacks. Hsu Kuang-ch'i, *Nung-cheng ch'üan-shu* (Peking, 1959 reprint), 31:10a. Ch'en Heng-li, p. 206. There were several variations on the straw stack idea. *TSYL*, 13b; *KTSSCP*, 19a.

43. *KTSSCP*, 2:21b; *Ch'ang-hsing HC* (1892), 8:24a–25a.

44. Sung Ying-hsing, p. 42.

45. *TSCMCS*, 20b–21a; *KTSSCP*, 2:3b–5a; and *Silk*, p. 54.

46. *HTS*, pp. 73–74; *Chia-hsing FC* (1879), 32:22b–23a.

47. *Survey*, pp. 2, 54–58, and 77–80.

48. *Kuei-an HC* (1892), 11:1a; *Chia-hsing HC* (1906), 15:13b–14a.

49. Tseng T'ung-ch'un, *Chung-kuo ssu-yeh* (Shanghai, 1933), p. 32; and D. K. Lieu, p. xiv.

50. *TSCMCS*, 17a–b.

51. *Chia-hsing FC* (1879), 32:39b; *KTSSCP*, 2:47b; Sung Ying-hsing, p. 38.

52. *Silk*, p. 126.

53. *ST*, pp. 127–132. The Japanese observers found that there was a great deal of regional variation in both superstitions and techniques. See ibid., pp. 34, 57–60. Also, *Survey*, p. 68; and *SYC: Che-chiang*, IV, 188–191. For examples of local variations in the Ch'ing, see *HTS*, p. 6; *TSCMCS*, 25b; and *KTSSCP*, 2:25b.

54. Mao Tun, *Spring Silkworms and Other Stories* (Peking, 1956), especially pp. 23–25.

55. *Survey*, pp. 8–9.

56. Tseng T'ung-ch'un, p. 162.

57. For example, see Tadao Yokoyama, pp. 196, 222.

58. *Wu-ch'ing CC* (1936), 7:18a–b.

59. *Survey*, p. 25. See also Table 3, where the yields are not so good, but represented non-test conditions.

60. *TSYL*, 14b; *TSCMCS*, 37a; *KTSSCP*, 2:27a–b; and *Silk*, p. 53.

61. Sung Ying-hsing, p. 48; *TSYL*, 15b; *TSCMCS*, 37b–38a; and *Silk*, p. 58, all have detailed descriptions of this process.

62. *HTS*, pp. 70–71. *Ch'ang-hsing HC* (1892), 8:32. *Nan-hsun CC* (1922), 32:20, has a description of various kinds of waste silk. Also, see *Wu-ch'ing CC* (1936), 21:10b.

63. *HTS*, pp. 66–69.

64. *Ch'ang-hsing HC* (1892), 8:25.

65. See, for example, *Nan-hsun CC* (1922), 30:22b; Hatano Yoshihiro, p. 22. Some localities still practiced it in modern times. See, for example, *TSCMCS*, 35b; *Wu-ch'ing CC* (1936), 7:13; Tseng T'ung-ch'un, p. 46.

66. E-tu Zen Sun, "Sericulture and Silk Textile Production in Ch'ing China," in W. E. Willmott, ed., *Economic Organization in Chinese Society* (Stanford, 1972), pp. 84–85.

67. Hsu Kuang-ch'i, 33:15b.

68. *KTSSCP*, 2:25b.

69. *Silk*, p. 53.

70. T. R. Banister, "A History of the External Trade of China, 1834–1881," in China, Maritime Customs, *Decennial Reports, 1922–1931* (Shanghai, 1933), I, 128. Also, A. J. Sargent, *Anglo-Chinese Commerce and Diplomacy* (Oxford, 1907), pp. 217–218.

71. Hsu Kuang-ch'i, 32:20a–21b, illustrates both methods.

72. *Silk*, p. 57; *Ch'ang-hsing HC* (1892), 8:32; and *Survey*, p. 67.

73. Illustrations are found in Hsu Kuang-ch'i, 32:18–19; Sung Ying-hsing, pp. 46–47; and also *Silk*, Fig. xxiii, following p. 68.

74. According to *Shun-te HC* (1929), 1:25b, the hand-reel was used in the Hsien-feng (1851–1861) and T'ung-chih (1862–1874) periods, while the treadle-reel was not used until the beginning of the Kuang-hsu period (1875–1908). Steam filatures were also introduced at that time.

75. The illustrations in *T'ien-kung k'ai-wu* cited above are variously marked north China and south China, while the same woodblock prints are labeled the opposite way in Hsu Kuang-ch'i's *Nung-cheng ch'üan-shu*. *Silk*, pp. 68–69, shows a small treadle machine.

76. See glossary for definition of silk terms. *Nan-hsun CC* (1922), 32:19b.

77. *TSYL*, 15a; *T'ung-hsiang HC* (1881), ts'e 7, wu-ch'an, 2a; *Ch'ang-hsing HC* (1892), 8:32a.

78. *Shuang-lin CC* (1917), 14:10b. See Ch. 3, pp. 77–78.

79. Yueh Ssu-ping, p. 39. *SZ*, XIII, 595ff, has a list of different types of domestically reeled silks. *SK*, p. 121 and pp. 183–184, has a list of Chinese and English terms for types of raw silk.

80. The Chinese cocoon-drying oven had a tray construction, while the more complex Japanese oven used a conveyer belt. Both types are illustrated in *Survey*, pp. 10–13.

81. D. K. Lieu, p. 136; and *Survey*, p. 14.

82. D. K. Lieu, pp. 113–117; Yueh Ssu-ping, pp. 92–93.

83. *Survey*, pp. 16–17; *ST*, p. 241; *SZ*, XV, 690–692; and Howard and Buswell, pp. 130–131.

84. *Survey*, p. 17.

85. For the early Ch'ing, see Ch'en Heng-li, p. 207; and Sung Ying-hsing, p. 49. For the late Ch'ing, see *Silk*, p. 57; *KTSSCP*, 2:26a; *TSCMCS*, 33a, 35a.

86. *SK*, p. 308, and *SZ*, XII, 55, in Shih Min-hsiung, *Ch'ing-tai ssu-chih kung-yeh ti fa-chan* (Taipei, 1968), p. 32, both state that one person could reel only 12–13 *liang* per day, but Yueh Ssu-ping, p. 40, states that one person could reel a catty, or 16 *liang*, a day. Comprehensive data collected in *SYC: Che-chiang*, VII, 46 and in *SKYSTL*, III, 694–695, suggest a very wide range, from 12–50 *liang* per machine. Each machine, however, may have had more than one operator.

87. D. K. Lieu, pp. 106, 118, and 136; and *ST*, pp. 274–275, both state that 11–12 *liang* a day was average. Yueh Ssu-ping, p. 102, gives a figure of 8–9 *liang*. *Survey*, p. 16, says 9–11 *liang* per day, depending on the denier of silk reeled.

88. See, for example, *Ta-kung pao* (Tientsin), 4/1917, in *KYSTL*, IVA, 172–173; Banister, I, 162.

89. Tseng T'ung-ch'un, p. 80, prefers 15:1, while D. K. Lieu, pp. 136, 138, prefers 18:1. See also Table 3.

90. Yueh Ssu-ping, pp. 126–127. See p. 77.

91. *SZ*, XV, 733–734.

92. See Sung Ying-hsing, p. 51, for illustrations.

93. For example, ibid., pp. 52–53.

94. *Nan-hsun CC* (1922), 32:20a.

95. *SKYSTL*, III, 389–391; *Hang-chou shih ching-chi tiao-ch'a*, ed. Chien-she wei-yuan hui tiao-ch'a Che-chiang ching-chi so (Hangchow, 1932), p. 155.

96. Yokoyama Suguru, "Shindai no toshi kinuorimonogyō no seisan keitai," *Shigaku kenkyū*, 104:75 (1968). For standard illustrations of looms, see Sung Ying-hsing, pp. 55–57. Also *KZ*, XII, 257–258. *Silk*, figures xxvi–xxxii, after p. 68, present some extremely fine drawings of many types of looms.

97. *DR, 1892–1901,* pp. 428–430; and *Silk,* pp. 60–61.

98. D. K. Lieu, pp. 163–164.

99. *SZ,* XV, 780–784. The Jacquard loom was introduced about the same time to the cotton-weaving industry, where it received an enthusiastic reception. Kang Chao, pp. 75–76.

100. D. K. Lieu, pp. 162–163.

101. I base this statement on a comparison of figures cited by Charles J. Huber, *The Raw Silk Industry of Japan* (New York, 1929), with data in Tables 1–3. According to Huber, p. 9, Japanese farmers in the 1920s were getting 370 *kamme* of leaves per *tan* per year, or 5,663 kg. per acre (at the rate of 1 *kamme* = 3.75 kg., and 1 *tan* = 0.2450 acre). If we say that the typical Chinese farmer was getting 1,000 catties of leaves per mou, he was obtaining a yield equivalent to 3,953 kg. per acre (at the rate of 1 catty = 0.6 kg., and 1 Ch'ing mou = 0.1518 acre), clearly a much inferior yield to the Japanese standard. See Dwight H. Perkins, *Agricultural Development in China, 1368–1968* (Chicago, 1969), Table G–1, p. 314. As Table 1 shows, however, the best results in the Ch'ing reached as high as 1,500–2,000 catties per mou, or 5,929 to 7,905 kg. per acre, which compares quite favorably with the twentieth-century Japanese results. Further, Huber states that it took 15 *kamme* of leaves to obtain 1 *kamme* of fresh cocoons—a result not significantly better than the average-to-better Chinese results, as shown in Table 2. Finally, on pp. 16–17, he states that the average Japanese yield of fresh cocoons to silk was 10:1, which was similar to the best Chinese results, as shown in Table 3.

102. Yagi Haruo, "Seishigyō," in *Nihon sangyōshi taikei,* comp. Chihōshi kenkyū kyōgikai (Tokyo, 1961), I, 236–238.

103. *Nung-shu,* Wang Chen (1924 ed.), 20:16b–20b, and *Nung-sang chi-yao (Wu-ying tien chü chen pan ch'üan-shu* ed.), 4:25a, are the earliest explicit references I have been able to find. The drawings in the former were apparently the basis for illustrations in the *T'ien-kung k'ai-wu* and Hsu Kuang-ch'i's *Nung-cheng ch'üan-shu.* Although *Fang-chih shih-hua,* comp. Shang-hai shih fang-chih k'o-hsueh yen-chiu yuan (Shanghai, 1978), pp. 57 and 233, states that the treadle reeling machine was introduced in the Sung, I personally have not found conclusive evidence for this. Neither Ch'en Fu's *Nung-shu,* nor Ch'in Kuan's *Ts'an-shu (Chih pu-tsu chai ts'ung-shu* eds., ts'e 69), two Sung manuals, mention it. Moreover, the *Yü-chih keng-chih t'u,* comp. Chiao Ping-chen (1696 ed.), an elaborate set of illustrations of agriculture and sericulture prepared for the K'ang-hsi Emperor in the late seventeenth century, and based on the Sung dynasty original edition of Lou Shou's *Keng-chih t'u,* does not show a treadle machine, but rather a reeling machine operated by means of a handle,

(see Figure 3). Lou Shou (1090–1162) was a native of the present-day Ningpo region.

Admittedly none of this evidence is conclusive, since perusal of later manuals reveals that illustrations of outdated technology were often reprinted even when they were not consistent with the text which they were intended to illustrate. A good example is *Ts'an-sang ts'ui-pien*, Wei Chieh (1956 Peking reprint of 1900 ed.), which on pp. 252–253 has some primitive drawings of hand-operated machines, while the text on pp. 130–131 gives a highly detailed description of the treadle machine. It is in fact quite likely that the treadle machine was already being used in the Sung, and that it was invented much earlier. *Fang-chih shih-hua*, p. 66, shows that a treadle spinning machine for cotton may have been in use in the fourth or fifth century.

104. This phenomenon of technological stagnation has already been described in general terms by Mark Elvin, *The Pattern of the Chinese Past* (Stanford, 1973), especially in Chapters 13, 14, and 17, and described, with special reference to agriculture, by Dwight H. Perkins, pp. 37–53.

105. Tadao Yokoyama, pp. 8–9. 1 *tan* = 0.2450 acre.

106. In 1924, Japan's output of steam filature silk was 412,000 piculs, and its output of hand-reeled silk was 27,500 piculs. Huber, p. 40. See Conclusion, n. 20 below.

107. Kang Chao, pp. 56–80, comes to a similar conclusion in his discussion of traditional cotton technology.

TWO   *The State and Traditional Enterprise*

1. Shiba Yoshinobu, pp. 113–121.

2. Ibid., pp. 111–113. Even in the Sung, silk weaving was done primarily for tax purposes and was closely supervised by officials. Yanagida Setsuko, "Sōdai no yōsan nōka keiei: Kōnan o chūshin to shite," in *Wada Hakushi koki kinen Tōyōshi ronsō*, ed. Wada Hakushi koki kinen Tōyōshi ronsō hensan iinkai (Tokyo, 1961), pp. 993–1003.

3. See, for example, Han Ta-ch'eng, "Ming-tai shang-p'in ching-chi ti fa-chan yü tzu-pen chu-i ti meng-ya," in *Ming-Ch'ing she-hui ching-chi hsing-tai ti yen-chiu*, comp. Chung-kuo jen-min ta-hsüeh, Chung-kuo li-shih chiao-yen shih (Shanghai, 1956), pp. 16–19, 48; and Li Chih-chin, "Lun ya-p'ien chan-cheng i-ch'ien Ch'ing-tai shang-yeh-hsing nung-yeh ti fa-chan," in ibid., pp. 288–290. See also "Ya-p'ien chan-cheng ch'ien-hsi

wo-kuo she-hui ching-chi ti kai-k'uang," in *Chung-kuo chin-tai kuo-min ching-chi shih chiang-i,* ed. Hu-pei ta-hsueh cheng-chih ching-chi-hsueh chiao-yen shih (Peking, 1958), pp. 33–40.

4. These views are summarized in Albert Feuerwerker, ed., *History in Communist China* (Cambridge, Mass., 1968), pp. 229–232.

5. Important articles were collected in *Chung-kuo tzu-pen chu-i meng-ya wen-t'i t'ao-lun chi,* comp. Chung-kuo jen-min ta-hsueh, Chung-kuo li-shih chiao-yen shih (Peking, 1957). A concise overview of this debate, with a full bibliography, is contained in Tanaka Masatoshi, "'Shihon shugi no hōga' kenkyū," in his *Chūgoku kindai keizaishi kenkyū josetsu* (Tokyo, 1973), pp. 205–241.

6. P'eng Tse-i, "Ch'ing-tai ch'ien-ch'i Chiang-nan chih-tsao ti yen-chiu," *Li-shih yen-chiu* 1963.4:91 (April 1963).

7. P'eng Tse-i, "Ts'ung Ming-tai kuan-ying chih-tsao ti ching-ying fang-shih k'an Chiang-nan ssu-chih-yeh sheng-ch'an ti hsing-chih," *Li-shih yen-chiu* 1963.2:34–35 (February 1963), especially note 10.

8. Saeki Yūichi, "Mindai shōekisei no hōkai to toshi kinuorimonogyō ryūtsūshijō no tenkai," *Tōyō bunka kenkyūjo kiyō* 10:375–376 (November 1956), gives a table showing when the local factories ceased operations.

9. Chu Ch'i-ch'ien, *Ssu-hsiu pi-chi* (1930), shang, 15–17; and *Kuang-hsu Ta-Ch'ing hui-tien shih-li,* chüan 1195, in *SKYSTL,* I, 70–73; and Shih Min-hsiung, p. 11.

10. P'eng Tse-i, "Ming-tai," p. 43; and *SKYSTL,* II, 5.

11. Accounts and records for the Imperial Silk Factories are presumably still preserved in China and, as access to such materials becomes more feasible, we should be able to construct a better picture of government-generated demand for silk. In 1935, a catalog of such records was published: *Ch'ing chiu-ch'ao ching-sheng pao-hsiao ts'e mu-lu* (Peking, 1935). Professor Susan Naquin of the University of Pennsylvania and other members of the Ming-Ch'ing history delegation, which visited the People's Republic of China in the summer of 1979, have reported that the records of the Imperial Household (Nei-wu fu) are housed at the Ming-Ch'ing Archives in Peking. The records of the Grand Secretariat (Nei-ko), also housed at the Archives, apparently contain record books from the Six Boards. The report of this delegation is in press.

12. Saeki Yūichi, "Mindai shōekisei," pp. 406–407, note 19, and his "Min zenpanki no kiko," *Tōyō bunka kenkyūjo kiyō* 8:168–169 (March 1956). See also Ray Huang, pp. 9–11.

13. P'eng Tse-i, "Ming-tai," p. 36. Saeki Yūichi, "Min zenpanki," p. 169, gives the same figures, but adds 280,000 bolts as silk tax commuted to silk fabric. Compare Table 5.

14. P'eng Tse-i, "Ming-tai," p. 55.

15. Saeki Yūichi, "Mindai shōekisei," pp. 407–408, gives a complete table of special orders for silk from 1425 to 1625.

16. P'eng Tse-i, "Ming-tai," pp. 34–35; and *SKYSTL*, I, 90.

17. Chu Ch'i-ch'ien, shang, p. 16, in *SKYSTL*, I, 100; and P'eng Tse-i, "Ch'ing-tai," p. 100.

18. This is an inference based on the knowledge that the Nanking and Soochow factories reported expenses of 115,000 taels each in 1708. Jonathan D. Spence, *Ts'ao Yin and the K'ang-hsi Emperor: Bondservant and Master* (New Haven, 1966), p. 95. Presumably the expenses of the Nanking factory were similar.

19. P'eng Tse-i, "Ch'ing-tai," p. 101.

20. Spence, pp. 88–89.

21. P'eng Tse-i, "Ch'ing-tai," p. 100; and *SKYSTL*, I, 78, 100.

22. P'eng Tse-i, "Ming-tai," pp. 37–38.

23. Spence, pp. 82–86, 95, 102.

24. Saeki Yūichi, "Mindai shōekisei," p. 373; and P'eng Tse-i, "Ming-tai," p. 34. See also Ray Huang's discussion of the role of silk in the Ming tax system, p. 136.

25. Saeki Yūichi, "Mindai shōekisei," p. 397. According to Mi Chu Wiens, "Socioeconomic Change during the Ming Dynasty in the Kiangnan Area," PhD dissertation (Harvard University, 1973), p. 131, this *chüan* was not paid directly, but was commuted to cash because it was considered too coarse for actual use.

26. *Hang-chou shih ching-chi tiao-ch'a*, 6:1.

27. *T'ung-hsiang HC* (1877), ts'e 6, ts'ao-yun, 7a–b; *Wu-ch'ing CC* (1936), 20:2a–b.

28. P'eng Tse-i, "Ch'ing-tai," pp. 110–111; and *SKYSTL*, II, 8–12, 96–97.

29. *SKYSTL*, I, 220; and Spence, p. 103.

30. 1851 Hu pu memorials cited by P'eng Tse-i, "Ch'ing-tai," p. 101.

31. The three factories were ordered to slow down their activities and use up their stores of silk. See *(Ch'in-ting) Ta-Ch'ing hui-tien shih-li* (Kuang-hsu, ed.), chüan 1190, Nei-wu fu, k'u-tsang.

32. Saeki Yūichi, "Mindai shōekisei," p. 396.

33. Ch'en Shih-ch'i, *Ming-tai kuan shou-kung-yeh ti yen-chiu* (Wuhan, 1958), pp. 6–7; and Fu I-ling, *Ming-tai Chiang-nan shih-min ching-chi shih-t'an* (Shanghai, 1963), pp. 81–82; and Ch'en Heng-li, p. 15.

34. *Wu-chiang HC*, chüan 18, in Fu I-ling, *Ming-Ch'ing shih-tai shang-jen chi shang-yeh tzu-pen* (Peking, 1956), p. 16. A variation of this passage is used to describe the rise of Chen-tse. See *Ch'ien-lung Chen-tse CC*, as quoted in Saeki Yūichi, "Mindai shōekisei," p. 396.

35. Saeki Yūichi, "Mindai shōekisei," pp. 394–395.

36. See Ch. 3, p. 67.

37. Saeki Yūichi, "Mindai shōekisei," pp. 386–393, has an excellent discussion of this functional differentiation.

38. Saeki Yūichi, "Min zenpanki," pp. 170–171, 175. For a discussion of other government industries, see Ch'en Shih-ch'i, pp. 42–45.

39. P'eng Tse-i, "Ming-tai," p. 39; and Saeki Yūichi, "Min zenpanki," pp. 179–180.

40. Ch'en Shih-ch'i, pp. 77–80.

41. Saeki Yūichi, "Min zenpanki," pp. 175, 182. Ch'en Shih-ch'i, pp. 71–77, explains how the *lun-pan chiang* system worked for government handicrafts in general. See also p. 83.

42. P'eng Tse-i, "Ming-tai," p. 38; and Saeki Yūichi, "Mindai shōekisei," p. 378.

43. Ibid., p. 363, and Saeki Yūichi, "Min zenpanki," pp. 188–189.

44. P'eng Tse-i, "Ming-tai," p. 49. Ping-ti Ho, *The Ladder of Success in Imperial China: Aspects of Social Mobility, 1368–1911* (New York, 1964), p. 57.

45. Saeki Yūichi, "Mindai shōekisei," pp. 364–365.

46. Ibid., pp. 372–376.

47. Ibid., p. 370.

48. P'eng Tse-i, "Ming-tai," pp. 43–46. P'eng notes that the *pao-lan* system was also employed to procure *chüan,* which were required to be paid as a form of tax by localities specializing in silk.

49. P'eng Tse-i, "Ch'ing-tai," pp. 92–94. Also, *Ch'ien-lung chung-hsiu Yuan-ho hsien-chih,* 10:17, in *SKYSTL,* I, 101.

50. P'eng Tse-i, "Ch'ing-tai," pp. 95–96; *SKYSTL,* I, 78.

51. P'eng Tse-i, "Ch'ing-tai," pp. 95–96, 112–114.

52. Ibid.

53. E-tu Zen Sun, p. 101.

54. *Ku-chin t'u-shu chi-ch'eng,* chüan 676, Su-chou, in Fu I-ling, *Ming-Ch'ing shih-tai shang-jen chi shang-yeh tzu-pen,* p. 12. See also E-tu Zen Sun's translation of this passage in "Sericulture and Silk Textile Production," p. 96. See this and similar passages from other sources in *SKYSTL,* I, 24.

55. Yokoyama Suguru, "Shindai," 104:77.

56. P'eng Tse-i, "Ming-tai," pp. 51–53, and "Ch'ing-tai," p. 112.

57. P'eng Tse-i, "Ya p'ien chan-cheng ch'ien Ch'ing-tai Su-chou ssu-chih-yeh sheng-ch'an kuan-hsi ti hsing-shih yü hsing-chih," *Ching-chi yen-chiu* 10:68–70 (October 1963). Also Sun P'ei, *Su-chou chih-tsao chü chih,* 10:7, in *SKYSTL,* I, 102, 214. See Ch. 4, pp. 127–129.

58. Saeki Yūichi, "Sen-roppyaku-ichi-nen 'shokuyō no hen' o meguru

shomondai," *Tōyō bunka kenkyūjo kiyō* 45:77–96 (March 1968); Fu I-ling, *Ming-tai Chiang-nan shih-min,* pp. 89–90; and P'eng Tse-i, "Ming-tai," pp. 52–53.

59. Liu Yung-ch'eng, "Shih-lun Ch'ing-tai Su-chou shou-kung-yeh hang-hui," *Li-shih yen-chiu* 1959.11:39–41 (November 1959). However, Liu reneged a bit by admitting that, since the owners often did not participate directly in production, the *chang-fang* also represented a feudal or semi-feudal form.

60. Yokoyama Suguru, "Shindai," 105:64–66.

61. Elvin, pp. 283–284.

62. P'eng Tse-i, "Ch'ing-tai," pp. 112–115, and "Su-chou ssu-chih-yeh," pp. 64–66.

63. Saeki Yūichi, "Shokuyō no hen," pp. 98–103.

64. See *SKYSTL,* II, 426 ff.

65. Although Hatano Yoshihiro, p. 36, says that the Soochow *chang-fang* did some weaving on their own premises, Yokoyama Suguru, "Shindai," 104:71–73, disagrees, saying that there is no evidence of this. It appears that the latter view is correct.

66. Kojima Yoshio, "Shin-matsu minkoku-shoki Soshūfu no kinuorigyō to kiko no dōkō," *Shakai keizai shigaku* 34.5:37–38 (January 1969).

67. Yokoyama Suguru, "Shindai," 105:55.

68. *SKYSTL,* III, 652.

69. Yokoyama Suguru, "Shindai," 105:53–54.

70. *SKYSTL,* III, 652–653; and *SYC: Chiang-su,* VIII, 167.

71. *SYC: Chiang-su,* VIII, 167.

72. P'eng Tse-i, "Su-chou ssu-chih-yeh," p. 65. Also, *SKYSTL,* I, 214, 410–411, and III, 219–220, 651ff. And Yokoyama Suguru, "Shindai," 104:74.

73. Yokoyama Suguru, "Shindai," 104:68; and P'eng Tse-i, "Ming-tai," p. 50.

74. Shih Min-hsiung, pp. 78–79, and *SKYSTL,* I, 222–223.

75. Ch'en Tso-lin, "Feng-lu hsiao-chih," in his *Chin-ling so-chih wu-chung* (1900), 3:2–5.

76. *DR, 1892–1901,* p. 429.

77. Yokoyama Suguru, "Shindai," 104:75.

78. With silk woven on power looms in the Republican period, dyeing would follow the weaving. Yokoyama Suguru, "Shindai," 105:62–63. See also, Ch. 1, n. 95.

79. Negishi Tadashi, *Shanhai no girudo* (Tokyo, 1951), p. 276.

80. *Shang-hai tsung-shang-hui yueh-pao,* 3/1924, 4:3, in *SKYSTL,* III, 219–220, or Li Han-ch'in, *Nan-ching tuan-yeh kai-k'uang,* 4.1:83–85 (1936), in *SKYSTL,* III, 652.

81. *Ch'ing-ch'ao hsu wen-hsien t'ung-k'ao,* chüan 385, shih-yen 8, in Yokoyama Suguru, "Shindai," 104:73.

82. *Chung-wai ching-chi chou-pao,* 10/1926, 186:14, in *SKYSTL,* III, 216–217.

83. *KZ,* XII, 268–272.

84. Negishi Tadashi, pp. 278–279; *KZ,* XII, 283.

85. Yokoyama Suguru, "Shindai," 104:73.

86. *Wu-ch'ing CC* (1936), 21:10.

87. *Kuo-chi mao-i tao-pao,* n.d., 4.5:36, in *SKYSTL,* II, 72–73.

88. *Ching-chi pan-yueh k'an,* 2.8, tiao-ch'a, pp. 18–19, 4/15/1928, in *SKYSTL,* III, 222–223; *SYC: Chiang-su,* VIII, 210–211, in *SKYSTL,* III, 721.

89. Yokoyama Suguru, "Shindai," 104:67–71, has a close review of the literature on this subject.

90. Eugene F. Rice, Jr., *The Foundations of Early Modern Europe, 1460–1559* (New York, 1970), pp. 47–53; and Maurice Dobb, *Studies in the Development of Capitalism,* rev. ed. (New York, 1963), pp. 7–8, 142–145.

91. Dobb, pp. 151–161.

92. Rice, p. 52; and G. D. Ramsay, *The Wiltshire Woollen Industry in the Sixteenth and Seventeenth Centuries,* 2nd ed. (London, 1965).

93. Rice, p. 151.

94. Ramsay, pp. 6–7,

95. For a general discussion, see Dobb, pp. 148–149, and Rice, pp. 52–53. I am grateful to Professor Robert Du Plessis, Department of History, Swarthmore College, for clarification of these issues and for suggesting sources for European history.

96. Dobb, p. 160.

97. The debate specifically concerns Dobb's interpretation of a passage from Chapter 20 of Part III of *Capital,* in which Marx discusses "the two ways" in which the feudal mode of production is transformed into a capitalistic mode of production. This debate originally appeared in the pages of *Science and Society* and other journals in the 1950s and has now been reprinted in *The Transition from Feudalism to Capitalism,* introd. by Rodney Hilton (London, 1976), pp. 52–55, 90–94, 100–101, and 137–140.

98. The reasons for the origin of the Industrial Revolution in England are still a source of academic controversy. Neil J. Smelser, *Social Change in the Industrial Revolution: An Application of Theory to the British Cotton Industry* (Chicago, 1959), pp. 56–68, 77ff., and 404, provides one provocative interpretation.

99. Whether there was a putting-out system in the Chinese cotton

industry has been a point of controversy among historians. The evidence for it is extremely fragmentary and is most fully discussed in Terada Takanobu, "Soshū chihō ni okeru toshi no mengyō shōnin ni tsuite," *Shirin* 41.6:52–69 (1958). Other evidence is discussed in Kitamura Hiranao, "Shindai ni okeru Koshūfu Nanjinchin no mentonya ni tsuite," *Keizaigaku zasshi,* 57.3:1–19 (September 1967). Kang Chao, pp. 53–55, also reviews some of the evidence, but regards the examples of putting out as very isolated.

100. Elvin, pp. 274–277.

101. Ibid., p. 283.

102. Yokoyama Suguru, "Tampugyō no seisan kōzō" in his *Chūgoku kindaika no keizai kōzō* (Tokyo, 1972), pp. 63–146.

103. Lien-sheng Yang, "Government Control of Urban Merchants in Traditional China," *Tsinghua Journal of Chinese Studies* 8:1 and 2:186ff. (August 1970). See also his "The Concept of 'Pao' as the Basis for Social Relations in China," in John K. Fairbank, ed., *Chinese Thought and Institutions* (Chicago, 1957), pp. 291–301, in which he discusses the functional relationship between this *pao* and two other characters also pronounced *pao,* the first meaning "security, guarantee of no failure," and the other meaning "reciprocity, retribution."

THREE   *The Silk Export Trade*

1. At least this was said to have been the case in the Roman Empire. Hudson, p. 66; and Ying-shih Yü, pp. 158–159.

2. Serruys, pp. 212–231.

3. Clifford M. Foust, *Muscovites and Mandarins: Russia's Trade with China and its Setting, 1727–1805* (Chapel Hill, 1969), pp. 112, 232, and 356.

4. Michael Cooper, "The Mechanics of the Macao-Nagasaki Silk Trade," *Monumenta Nipponica* 27.4:423–425 (Winter 1972).

5. C. R. Boxer, *The Great Ship from Amacon: Annals of Macao and the Old Japan Trade, 1555–1640* (Lisbon, 1963), p. 39.

6. Ibid., pp. 39, 179. A ducat was roughly equivalent to one Chinese silver tael.

7. Fujimoto Jitsuya, *Nihon sanshigyōshi* (n.p., 1933), III, 34–35, 55. A picul equaled 100 catties. See also Conclusion.

8. Ibid., I, 148.

9. A thorough discussion of this trade is found in William Lytle Schurz, *The Manila Galleon* (New York, 1939). See especially p. 32, which

describes the composition of the trade. I am grateful to Louisa Hoberman of the Institute for Latin American Studies, University of Texas, Austin, for referring me to this source, and for clarification of the nature of the Manila-Mexico trade.

10. Ch'üan Han-sheng, "Tzu Ming-chi chih Ch'ing chung-yeh Hsi-shu Mei-chou ti Chung-kuo ssu-huo mao-i," in his *Chung-kuo ching-chi shih yen-chiu* (Hong Kong, 1976), I, 460–469.

11. Ibid., pp. 459–460, 465–466. Boxer, p. 336. 1 peso = 0.7–0.9 Chinese taels. Consequently, 2 to 3 million pesos was the equivalent of 1.6–2.4 million taels, or at least 3,200–4,800 piculs of raw silk at the rate of 500 taels per picul. Since this price was extraordinarily high, the actual quantity of silk imports was probably much greater than 5,000 piculs. The second calculation is based on the equivalent of 80 catties per bale, which was the standard in the nineteenth century. See also Schurz, pp. 59 and 189. My calculation is based on raw silk. In fact, fabrics constituted a substantial part of the trade, although the exact proportions of raw silk and woven silk are unknown. Since the value of fabrics was much higher than that of raw silk, the actual amount of silk, as measured by weight instead of value, was no doubt less than my rough calculations suggest.

12. Ch'üan Han-sheng, "Mei-chou pai-yin yü shih-pa shih-chi, Chung-kuo wu-chia ko-ming ti kuan-hsi," *Chung-yang yen-chiu yuan li-shih yü-yen yen-chiu so ch'i-k'an* 28:517–550 (1957). See also William S. Atwell, "Notes on Silver, Foreign Trade, and the Late Ming Economy," *Ch'ing-shih wen-t'i* 3.8:1–33 (December 1977).

13. John E. Wills, Jr., "Ch'ing Relations with the Dutch, 1662–1690," in John King Fairbank, ed., *The Chinese World Order: Traditional China's Foreign Relations* (Cambridge, Mass., 1968), pp. 236, 244; and L. Dermigny, *La Chine et l'occident: le commerce à Canton au XVIIIe siècle, 1719–1833* (Paris, 1964), I, 393.

14. Earl H. Pritchard, *The Crucial Years of Early Anglo-Chinese Relations, 1750–1800* (Pullman, Washington, 1936), p. 166; and Hosea Ballou Morse, *The Chronicles of the East India Company Trading to China, 1635–1834* (Oxford, 1926), I, 39–40, 44–46.

15. Pritchard, pp. 146, 167. Yen Chung-p'ing, et al., comp., *Chung-kuo chin-tai ching-chi shih t'ung-chi tzu-liao hsuan-chi* (Peking, 1955), p. 14, has a table showing the values of tea and silk exports between 1760 and 1833.

16. Morse, *Chronicles*, V, 31.

17. Hosea Ballou Morse, *The International Relations of the Chinese Empire* (London, 1910–1918), I, 168, 366. 1 bale = 0.8 picul or 80 catties.

18. M. Dorothy George, *England in Transition*, rev. ed. (Baltimore, 1953), p. 103.

19. William Milburn, *Oriental Commerce* (London, 1813), II, 249–251; Morse, *Chronicles*, V, 31 and II, 137; and Dermigny, I, 403.

20. Milburn, II, 251–256.

21. Morse, *Chronicles*, II, 89–96, 126 and V, 30–31.

22. Morse, *Chronicles*, V, 69.

23. *Ta-Ch'ing li-ch'ao shih-lu*, 591:13b–14a, as translated in Lo-shu Fu, comp. and trans., *A Documentary Chronicle of Sino-Western Relations* (Tucson, 1966), p. 226.

24. *Ta-Ch'ing li-ch'ao shih-lu*, 660:13a–b, as translated in Lo-shu Fu, p. 231.

25. *Ta-Ch'ing li-ch'ao shih-lu*, 704:10a–11b, 707:5a–6b, 708:7a–b, as translated in Lo-shu Fu, pp. 235–236.

26. Morse, *Chronicles*, II, 96. Twenty-three ships a year were permitted. *Hu pu tse-li* (1865 ed.), 41:28b.

27. Dermigny, I, 405.

28. Morse, *Chronicles*, II, 96; III, 181; and II, 54. On prices, see also E-tu Zen Sun, "Sericulture and Silk Production," pp. 90–91; Spence, pp. 100–101, 295–296; and Shih Min-hsiung, pp. 21–22.

29. Yeh-chien Wang, "The Secular Trend of Prices during the Ch'ing Period (1644–1911)," *Journal of the Institute of Chinese Studies of the University of Hong Kong* 5.2:351 (1972).

30. Ch'üan Han-sheng, "Mei-chou pai-yin," pp. 517–550.

31. Jean Baptiste Du Halde, *The General History of China* (Paris, 1735), II, 356.

32. Banister, p. 22.

33. S. Wells Williams, *The Chinese Commercial Guide*, 5th ed. (Hong Kong, 1863), p. 137.

34. Fortune, *Two Visits to the Tea Countries of China*, II, 12.

35. Augustine Heard and Co., Archives (Baker Library, Harvard Business School), Prices Current, Shanghai, September, December, 1857, and January, 1858.

36. Morse, *International Relations*, I, 358; and Rhoads Murphey, *Shanghai: Key to Modern China* (Cambridge, 1953), p. 103.

37. Banister, p. 49. See John K. Fairbank, *Trade and Diplomacy on the China Coast: Opening of the Treaty Ports, 1842–1854* (Cambridge, Mass., 1953), pp. 300–302, on the problem of transit tariffs on silk between the two treaty settlements.

38. Murphey, *Shanghai*, p. 70. Yen-p'ing Hao, *The Compradore in Nineteenth Century China* (Cambridge, Mass., 1970), pp. 53, 76, and 82.

39. C. F. Remer, *The Foreign Trade of China* (Shanghai, 1926), pp. 42–45.

40. Since it is generally held that the Customs Service undervalued

exports, the actual increases in the value of silk exports may have been greater than these figures would suggest. Remer, p. 181; and Tuan-liu Yang and Hou-pei Hou, *Statistics of China's Foreign Trade during the Last Sixty-five Years, 1864–1928* (Nanking, 1931), p. v.

41. *DR, 1922–1931,* I, 190, has annual figures for 1882–1931.

42. *Survey,* p. 83.

43. Banister, p. 99. It should be noted that in the earlier years a great portion of Canton's exports probably were unrecorded, because they were shipped by junk to Macao and Hong Kong rather than to Europe. Banister, p. 127. These figures were not usually included in the Customs returns until at least 1887. Remer, p. 44. In the twentieth century, the Canton weaving industry became technologically backward; most of its output was for the domestic market. Howard and Buswell, p. 155.

44. Ralph E. Buchanan, *The Shanghai Raw Silk Market* (New York, 1929), pp. 13–15, has a list of categories of silk exports.

45. See for example, *ATRR 1889,* p. 187, and also Ch. 1.

46. *AR 1907,* p. 54; *AR 1910,* p. 22. Also, *ATRR 1912,* p. 453; *ATRR 1923,* p. 8; Remer, pp. 136–137.

47. *Silk: Replies from the Commissioners of Customs, plus "Manchurian Tussore Silk,"* comp. China, Maritime Customs (Shanghai, 1917), pp. 184–185.

48. Liang-lin Hsiao, *China's Foreign Trade Statistics, 1864–1949* (Cambridge, Mass., 1974), p. 114.

49. Banister, p. 108. See Liang-lin Hsiao, pp. 109–110, for annual exports of cocoons and waste products by weight and value.

50. Buchanan, pp. 15–18. See also *ST,* pp. 455–485, for a detailed discussion of the waste silk trade.

51. S. Wells Williams, *The Middle Kingdom,* rev. ed. (New York, 1883), II, 395.

52. Banister, pp. 65, 69.

53. Dermigny, I, 402–404.

54. Louis Pasteur, *Etudes sur la maladie des vers à soie* (Paris, 1870), I, 3–4. 1 picul = 60 kilograms.

55. "Canton Exports of Raw Silk and Silk Waste in 1927," *Chinese Economic Journal* 2.6:529–532 (June 1928). In 1923, Canton exported 37,874 piculs of raw silk to the United States and 13,926 piculs to France (calculated at 80 catties per bale), while in 1927, Canton shipped 30,230 piculs to the United States and 22,534 to France. Howard and Buswell, p. 144, suggest that this may have been due to the deteriorating standards by which Canton silk was reeled. The French could use lower grades of silk. "Although French orders pay a lower price for raw silk, many filatures prefer to take only orders for France because of the greater ease in

carrying out the work. Most filatures which reel for both countries put their best women onto American orders, and of course rereel for these orders."

56. *AR 1900*, p. 32; *AR 1908*, 23-24.

57. *AR 1917*, p. 19. "Shanghai Silk Filatures," *Chinese Economic Journal* 3.1:590 (July 1928).

58. *AR 1882*, p. 10; and *AR 1883*, p. 11.

59. *AR 1875*, pp. 20-21; and *AR 1876*, pp. 30-31.

60. H. D. Fong, "China's Silk Reeling Industry: A Survey of Its Development and Distribution," *Monthly Bulletin on Economic China* 7.12:487 (December 1934).

61. Keishi Ohara, comp. and ed., *Japanese Trade and Industry in the Meiji-Taisho Era* (Tokyo, 1957), pp. 253-260.

62. Remer, p. 140.

63. D. K. Lieu, p. xv; *She-hui tsa-chih*, 2/1931, in KYSTL, IVA, 128.

64. *AR 1906*, p. 69.

65. *AR 1913*, p. 36.

66. *AR 1914*, p. 15.

67. *AR 1919*, p. 706.

68. Jesse W. Markham, *Competition in the Rayon Industry* (Cambridge, Mass., 1952), p. 33.

69. Banister, p. 128.

70. *ATRR 1905*, p. 238.

71. D. K. Lieu, pp. xiv-xv; *ST*, pp. 447-449.

72. Banister, p. 126.

73. *AR 1911*, p. 38.

74. *ATRR 1883*, p. 164; *DR, 1882-1891*, p. 317; and *ATRR 1884*, p. 162; and C. John Stanley, *Late Ch'ing Finance: Hu Kuang-yung as an Innovator* (Cambridge, Mass., 1961), pp. 73-76.

75. I am grateful to Professor Yeh-chien Wang of Kent State University for pointing this out to me.

76. Remer, pp. 78-79, 127.

77. Buchanan, pp. 38-39. On the relationship between foreign trade and currency speculation, see Andrea Lee McElderry, *Shanghai Old-Style Banks (Ch'ien-chuang), 1800-1935* (Ann Arbor, 1976), p. 20.

78. *ATRR 1906*, p. 234.

79. Remer, pp. 38-41.

80. Banister, pp. 70, 73.

81. *DR, 1882-1891*, p. 323.

82. Tseng T'ung-ch'un, pp. 96-97; Buchanan, pp. 25-27; and *ST*, pp. 426-428, has a list of these firms. In 1929 there were only four Chinese raw silk export companies at Shanghai—the Tonying Silk Trading Company,

the Hu Lin, the Wei Chen, and the Hwa Tung. By 1932 only the Tonying Company was still in business. D. K. Lieu, p. 144. See Chapter 6.

83. Buchanan, p. 22.

84. D. K. Lieu, p. 143.

85. Buchanan, pp. 27–28.

86. *ST,* p. 429.

87. Buchanan, p. 23.

88. *ST,* pp. 429–431; and Buchanan, p. 23.

89. The rate varied according to the prestige of the broker, but 3.75 taels per picul of filature silk, or 3.90 taels per picul of Tsatlee silk, was not uncommon. Buchanan, pp. 18–23.

90. Capitalization of brokerages was usually at 100,000 yuan or more. D. K. Lieu, p. 142.

91. *ST,* p. 431.

92. *Wu-hsing nung-ts'un ching-chi,* comp. Chung-kuo ching-chi t'ung-chi yen-chiu so (Shanghai, 1939), p. 216; also in *SKYSTL,* II, 83.

93. *Mushaku no seishigyō,* comp. Minami Manshū tetsudō (Shanghai, 1940), p. 5. See also Chapter 6.

94. Buchanan, p. 42.

95. *ST,* pp. 395–398.

96. "The Silk Export Trade of China," *Chinese Economic Journal* 3.5:954 (November 1928); Howard and Buswell, pp. 147–149.

97. Robert Eng makes this point very cogently in his PhD dissertation, "Imperialism and the Chinese Economy: The Canton and Shanghai Silk Industry, 1861-1932" (University of California, Berkeley, 1978), pp. 116–121.

98. Banister, p. 125, referring specifically to the tea guilds before 1884.

99. Albert Feuerwerker, *The Chinese Economy, ca. 1870–1911* (Ann Arbor, 1969), pp. 56–58.

100. This is the central idea of Murphey's *The Treaty Ports and China's Modernization: What Went Wrong?* and the title of Chapter 7.

FOUR   *Foreign Trade and Domestic Growth*

1. *Silk.*

2. Natalis Rondot, *Les soies,* 2nd ed. (Paris, 1885–1887). See Table 17.

3. These were reprinted in 1912 under the auspices of a Japanese group in Tientsin. *Shina seisan jigyō tōkeihyō,* comp. Shinkoku nōkōshō-bu (Osaka, 1912). See Table 20.

4. *Nung-shang t'ung-chi piao,* comp. Chung-kuo nung-shang pu (Peking, 1914–1924).

5. *Wu-ch'ing CC* (1936), 21:19b.

6. *SK,* pp. 5–6; and *ST,* p. 23.

7. Howard and Buswell, p. 38.

8. H. D. Fong, p. 484.

9. Kawabata Genji, "Taihei tenkoku senryōka Nanjunchin ni okeru Koshi bōeki," *Tōhōgaku* 22:95 (July 1961).

10. *Nan-hsun CC* (1922), 45:6ff.

11. Kawabata Genji, pp. 87–90.

12. *Ch'ang-hsing HC* (1892), 8:la.

13. *Silk,* p. 80.

14. *Wu-hsing nung-ts'un ching-chi,* p. 211; also quoted in *NYSTL,* I, 392.

15. Banister, p. 128.

16. *Silk,* p. 71.

17. Ibid., p. 79.

18. *Nan-hsun CC* (1922), 30:21b.

19. *HTS,* pp. 66–67.

20. *An-chi HC* (1875), 8:29, in *NYSTL,* I, 428.

21. *P'ing-hu HC* (1886), 2:51b, and also 8:36, in *NYSTL,* I, 429.

22. *Chia-hsing FC* (1879), 32:22a–b; Ch'en Heng-li, pp. 17, 179.

23. *Shih-men HC* (1878), 11:4a.

24. *Chia-hsing hsien nung-ts'un tiao-ch'a,* ed. Feng Tz'u-kang (Hangchow, 1936), pp. 70–71, says that between 1911 and 1935 the amount of cocoons sold to local cocoon hongs increased from 3,184 piculs annually to 14,135 piculs. When these figures are compared with Table 19, it can be inferred that the portion reeled domestically was still by far the larger.

25. *Nung-shang kung-pao,* 10/1934, in *NYSTL,* II, 257; *Che-chiang ching-chi lueh,* pp. 3–4, in *NYSTL,* II, 192.

26. *Wu-hsing nung-ts'un ching-chi,* p. 10, reported that only 185,169 mou were still planted in mulberry, whereas previously as much as 545,569 mou had been devoted to mulberry. See *SYC: Che-chiang,* IV, 167–168, 184–185.

27. *Che-chiang sheng nung-ts'un tiao-ch'a,* ed. Chung-kuo hsing-cheng yuan, nung-ts'un fu-hsing wei-yuan hui (Shanghai, 1934), p. 5.

28. Ibid., p. 125.

29. For example, *Te-ch'ing HC* (1931), 4:38a, 3a. The gazetteer said that Te-ch'ing had always needed to import half its annual supply of rice but, since the price of rice had risen from 2 to 3 yuan per picul to 10 or more in 1920, the suffering of the people had become great. Those who were not poor before were poor now, and those who were only slightly poor before were now very poor, it said.

30. *Silk*, pp. 82, 111.

31. *Chu-chi HC* (1903), 20:19b.

32. Yueh Ssu-ping, pp. 175–178; and *Survey*, p. 75.

33. *Hang-chou FC* (1784), 53:7b.

34. *Silk*, p. 80.

35. *Che-chiang ching-chi lueh*, p. 2; Hai-ning hsien, in *NYSTL*, II, 192; and *Chung-wai ching-chi chou-k'an*, 6/1927, in *NYSTL*, II, 223.

36. Yeh-chien Wang, "The Impact of the Taiping Rebellion on the Population in Southern Kiangsu," *Papers on China* 19:121–122, 131–139 (December 1965).

37. *Silk*, p. 62.

38. Yueh Ssu-ping, pp. 118–119.

39. *Chü-jung HC* (1904), 4:12–13, in *NYSTL*, I, 427. Tso had also started a sericultural bureau when he was governor-general of Chekiang and Fukien in 1863–1866. Arthur W. Hummel, ed., *Eminent Chinese of the Ch'ing Period* (Washington, D.C., 1943–1944), p. 764.

40. *Tung-fang tsa-chih*, 5/1905, in *NYSTL*, I, 884.

41. *Silk*, p. 59.

42. *Silk*, p. 61; and *Tan-yang HC* (1885), 29:7, in *NYSTL*, I, 427.

43. See Chapter 5. Also, *Tan-t'u HC* (1879), 17:19, in *NYSTL*, I, 426 and 883.

44. In 1914, 81% of its households were engaged in sericulture. *Shina seisan jigyō tokeihyō*, I, 703–704.

45. *Silk*, p. 61.

46. *Wu-hsi Chin-kuei HC* (1881), 31:1b.

47. *Silk*, p. 62, converted from 3,200,000 taels weight at 16 taels, or ounces, per catty.

48. *Ts'an-wu shuo-lueh*, ed. K'ang Fa-ta (F. Kleinwachter), n.d. in *Chih-hseuh ts'ung-shu ch'u-chi*, ts'e 27, 8b.

49. Chiao Lung-hua, *Nung-ts'un ching-chi*, 10/1934, in *NYSTL*, II, 190.

50. *Kōsoshō Mushakuken nōson jittai chōsa hōkokusho*, comp. Minami Manshū tetsudō kabushiki kaisha, Shanhai jimusho chōsashitsu (Shanghai, 1941), pp. 9–11; *Nung-shang kung-pao*, 8/1921, in *NYSTL*, II, 151.

51. *Survey*, pp. 29, 32.

52. Yueh Ssu-ping, pp. 298, 301.

53. *ST*, pp. 11–12. See also Table 19.

54. E-tu Zen Sun, "Sericulture and Silk Textile Production," p. 92; and *Shun-te HC* (1929), 1:24–26. These estimates are higher than those in Table 8.

55. Banister, pp. 127–128; and *ATRR 1877*, p. 214.

56. *Silk*, p. 148. At 13:1, this would mean the equivalent of 598,000 piculs of fresh cocoons; at 11:1, the equivalent of 552,000 piculs.

57. *DR, 1892–1901*, I, 177.

58. Suzuki Tomoo, "Shimmatsu-Minsho ni okeru minzoku shihon no tenkai katei: Kanton no seishigyō ni tsuite," in *Chūgoku kindaika no shakai kōzō,* in *Tōyō shigaku ronshū* (Tokyo, 1960), VI, 63.

59. *ST,* pp. 15, 1038. But estimates vary widely. Howard and Buswell, p. 155, estimated that 62% of south China's raw silk was used in local weaving and consumed domestically.

60. See Table 12. In 1899–1900, 36,987 piculs of raw silk were shipped from Canton, of which 34,612 were steam filatures. *DR, 1892–1901,* p. 177.

61. *ST,* p. 928.

62. Shiba Sakuo, "Shimmatsu Kanton sankakusu no yōsan keiei to nōson kindaika—Tōyōteki shakai to 'gyotō'," *Shikan,* nos. 57–58 (March 1960), 22–34. Also Morita Akira, *Shindai suirishi kenkyū* (Tokyo, 1974), pp. 141–170; and Eng, "Imperialism and the Chinese Economy," p. 138.

63. See Table 26, Ch. 5.

64. *ST,* p. 734; and Yueh Ssu-ping, pp. 277–278.

65. H. D. Fong, pp. 503–504.

66. Liang-lin Hsiao, p. 107.

67. *ST,* pp. 13–14, 736, 867. According to D. K. Lieu, p. xiii, only 30% of Szechwan silk was exported.

68. Yueh Ssu-ping, pp. 263–264; *ST,* p. 854.

69. *ST,* p. 769.

70. *ST,* p. 541.

71. H. D. Fong, p. 505; and Yueh Ssu-ping, p. 220.

72. *ST,* pp. 540, 555–556.

73. Liang-lin Hsiao, pp. 107, 113–114.

74. Yueh Ssu-ping, pp. 311, 341.

75. H. D. Fong, pp. 498–502; Yueh Ssu-ping, p. 629.

76. Yueh Ssu-ping, p. 309.

77. *ST,* pp. 13, 691.

78. Ch'en Tso-lin, 3:2a–b. *Silk,* pp. 62–63, states that in 1853 there were 35,000 looms in the city and 15,000 in the neighboring villages. Various gazetteer sources cited in *SKYSTL,* I, 215, 222–223, say that there were at least 30,000 looms in the Tao-kuang period (1821–1850) and before. *SZ,* XV, 785, states that there were at least 30,000 satin looms and 10,000 others in the 1820s.

79. *Silk,* 1917 ed., pp. 63–64, in *SKYSTL,* II, 67, says that there were 14,510 looms. Ch'en Tso-lin, 3:2a–b, says that there were 17,000–18,000 looms.

80. *SKYSTL,* I, 602; *SKYSTL,* III, 75; and Ch'en Tso-lin, 3:7a–10b.

81. Ch'en Tso-lin, 3:5a; and *SKYSTL,* II, 64.

82. *Silk,* pp. 62–64; *DR, 1892–1901,* p. 429.

83. *1900 Nan-ching k'ou hua-yang mao-i ch'ing-hsing lun-lueh,* hsia, p. 34, and *Chung-wai jih-pao,* 8/1900, both in *SKYSTL,* III, 596.

84. *Hsu-tsuan Chiang-ning FC*, chüan 15, in Yokoyama Suguru, "Shindai," 104:72; and also *SKYSTL*, III, 8–9, where other sources state that there were as many as 12–13,000 looms. *Kung-shang pan-yueh k'an*, 3:17, tiao-ch'a, p. 23, in *SKYSTL*, II, 452.

85. Yang Ta-chin, *Hsien-tai Chung-kuo shih-yeh chih* (Changsha, 1938), I, 146; and *SYC: Chiang-su*, VIII, 166.

86. *SYC: Chiang-su*, VIII, 167–172.

87. *Ch'ien-yeh yueh-pao*, 10/1925, in *SKYSTL*, III, 8–9, 220, 394.

88. *SZ*, XV, 788; *DR, 1922–1931*, p. 625; Yang Ta-chin, I, 146; and *SYC: Chiang-su*, VIII, 165.

89. *Kuang-hsu Chiang-ning FC*, in *SKYSTL*, II, 66–67. Also, *DR, 1892–1901*, p. 428.

90. Saeki Yūichi, "'Shokuyō no hen,'" p. 99.

91. *Silk*, pp. 73–74.

92. *Kuang-hsu pan-yueh-k'an*, 7:12, p. 48, in *SKYSTL*, II, 69; and *SYC: Chiang-su*, VIII, 222.

93. Yang Ta-chin, I, 148–149; *KZ*, XII, 259.

94. *SYC: Chiang-su*, VIII, 222; *SKYSTL*, II, 452, 428; and *DR, 1912–1921*, II, 66.

95. *KZ*, XII, 248–249.

96. *Shih-pao*, 8/1912, in Kojima Yoshio, p. 35.

97. *Silk*, pp. 70, 81.

98. *DR, 1902–1911*, p. 49.

99. *KZ*, XII, 298.

100. *KYSTL*, IVA, 103. See lists of types of silks in *SKYSTL*, III, 83; and *Hang-chou shih ching-chi tiao-ch'a*, 6:25ff.

101. D. K. Lieu, pp. 163–164.

102. *Hang-chou shih she-hui ching-chi t'ung-chi kai-yao*, pp. 43–44, in *SKYSTL*, III, 84.

103. *Hang-chou shih ching-chi tiao-ch'a*, 6:26–27, 53; *SYC: Che-chiang*, VII, 54. Also *Kuo-chi mao-i tao-pao*, 5/1933, in *SKYSTL*, III, 388–389.

104. *Hang-chou shih ching-chi tiao-ch'a*, 5:5–6.

105. *Survey*, p. 49.

106. *Hang-chou shih ching-chi tiao-ch'a*, 6:17–18, 22–25.

107. Yang Ta-chin, I, 148–149; *KZ*, XII, 259.

108. *SZ*, XV, 780–784.

109. D. K. Lieu, pp. 163–164.

110. *Kuo-chi lao-kung t'ung-hsun*, 5/1936, in *SKYSTL*, III, 429.

111. *Hang-chou shih ching-chi tiao-ch'a*, 6:62, 70.

112. *Chung-wai ching-chi chou-pao*, 10/1926, in *SKYSTL*, III, 4–5.

113. *SYC: Che-chiang*, VII, 52; and *SKYSTL*, III, 5; *Hang-chou shih ching-chi tiao-ch'a*, 6:2–3.

114. *Kuo-chi mao-i tao-pao*, 5/1933, in *SKYSTL*, III, 389.

115. *Kuo-chi lao-kung t'ung-hsun,* 5/1936, in *SKYSTL,* III, 429.

116. D. K. Lieu, pp. 161–165. See *SYC: Chiang-su,* VIII, 132–160, for details about the names, organization, and financing of the Shanghai weaving mills.

117. *Silk,* pp. 70, 79.

118. *Kuo-chi mao-i tao-pao,* 10/1932, in *SKYSTL,* III, 768–769.

119. *Ching-chi pan-yueh-k'an,* 4/1928, in *SKYSTL,* III, 86–87.

120. *Ts'an-ssu kai-liang shih-yeh kung-tso pao-kao,* 1934, in *SKYSTL,* III, 429–430.

121. *Kuei-an HC* (1882), 13:18a–b. Kuei-an and Wu-ch'eng were virtually part of the prefectural city of Hu-chou. See map in *Hu-chou FC* (1874), 1:1b–2a. After 1912, the two were combined to form Wu-hsing hsien, while Hu-chou city was renamed Wu-hsing.

122. *Te-ch'ing HC* (1923), 2:16a.

123. *Shuang-lin HC* (1917), 17:3.

124. Feng Tzu-tsai, *Shih-yeh t'ung-chi,* in *SKYSTL,* III, 85.

125. *SYC: Che-chiang,* III, 79.

126. *SYC: Che-chiang,* III, 80, and VII, 48.

127. *Silk,* p. 77. These did not necessarily operate all year around. D. K. Lieu, p. 13. See also Feng Tzu-tsai, *Shih-yeh t'ung-chi,* in *SKYSTL,* III, 85.

128. *SYC: Che-chiang,* III, 80, and VII, 48. See VII, 48–52, for a table listing all the factories in the locality.

129. *SYC: Chiang-su,* VIII, 215, in *SKYSTL,* II, 71–72; and *Nung-yeh chou-pao,* 7/1931, in *NYSTL,* III, 654.

130. *Chiang-su sheng-chien,* in *SKYSTL,* III, 86; and *SYC: Chiang-su,* VIII, 215–217.

131. Ch'ien-lung *P'u-chen chi-wen,* and Kuang-hsu *T'ung-hsiang HC,* in *SKYSTL,* I, 216–218.

132. *Silk,* p. 80.

133. *Silk,* p. 82, says it experienced a brief revival after the Taiping Rebellion, but it declined thereafter. Wang T'ing-feng, *Shao-hsing chih ssu-ch'ou,* in *SKYSTL,* II, 77.

134. *Silk,* p. 60; *DR, 1892–1901,* p. 444; *SKYSTL,* II, 428–431; and *KZ,* XII, 287.

135. *ST,* p. 24. *Survey,* p. 95, gives the following estimates: 20,000 at Hangchow, 15,000 at Hu-chou, 8,000 at Sheng-tse, 4,000 at P'u-yuan, 3,000 at Soochow, 2,000 at Nanking, and 5,000 at Shao-hsing, for a total of 57,000 piculs.

136. Shanghai International Testing House, *China Silk* (Shanghai, 1925), p. 8. Also, see Ch. 6, p. 181.

137. See for example, Liu Yung-ch'eng, pp. 30–31; and P'eng Tse-i, "Ming-tai," p. 50.

138. See examples in *Chiang-su sheng Ming-Ch'ing i-lai pei-k'o tzu-liao hsuan-chi,* comp. Chiang-su po-wu kuan (Peking, 1959), p. 2, introduction. and pp. 18–19. Also, P'eng Tse-i, "Su-chou ssu-chih-yeh," p. 69, and his "Shih-chiu shih-chi hou-ch'i, Chung-kuo ch'eng-shih shou-kung-yeh shang-yeh hang-hui ti ch'ung-chien ho tso-yung," *Li-shih yen-chiu* 1965.1:76 (January 1965). Also Yokoyama Suguru, "Shindai," 105:57–58.

139. P'eng Tse-i, "Hang-hui," pp. 72–73.

140. Negishi Tadashi, pp. 256, 261–264, 271–274. See *KZ,* XII, 316–317, for a complete list of Hangchow silk guilds.

141. Kojima Yoshio, pp. 39–51. *SKYSTL,* II, 597–600, has a complete list. On the 1910 crisis, see Marie-Claire Bergère, *Une crise financière à Shanghai à la fin de l'ancien régime* (Paris, 1964), especially pp. 5–7.

142. P'eng Tse-i, "Hang-hui," p. 91. In his "Su-chou ssu-chih-yeh," pp. 67ff, P'eng stresses the continued domination of the guilds over the industry.

FIVE   *Foreign Trade and the Rural Economy*

1. Fortune, *A Residence,* pp. 353–354.

2. *Wu-hsing nung-ts'un ching-chi,* pp. 122, 212, 220.

3. Yen Chung-p'ing, "Ming-Ch'ing liang-tai ti-fang kuan ch'ang-tao fang-chih yeh shih-li," *Tung-fang tsa-chih* 42.8:20–26 (April 1946).

4. *Ming Shih-lu* in *HTS,* p. 19; Ch'üan Han-sheng, "Sung-Ming chien pai-yin kou-mai-li ti pien-tung chi ch'i yuan-yin," *Hsin-ya hsueh-pao* 8.1:170 (February 1967), n. 6.

5. Ch'en Heng-li, p. 26.

6. *PNS,* pp. 27–28; Ch'en Heng-li, pp. 252–253.

7. Ibid. This passage is also in *Chia-hsing FC* (1879), 32:23b–24a.

8. Ch'en Heng-li, pp. 48–49, 252–253, gives a rather elaborate explication of Chang Lü-hsiang's passage.

9. Ibid., p. 51.

10. Ibid., p. 17. Also quoted in *Chia-hsing FC* (1879), 32:23a.

11. See various examples in *SKYSTL,* I, 203–208; and also in Yen Chung-p'ing, "Ming-Ch'ing," pp. 21–23.

12. *Tan-t'u HC* (1879), 17:19, in *NYSTL,* I, 426, 886.

13. *Ts'an-sang shuo* was supposed to have been written by a certain Shen Lien, based on Shen Ping-ch'eng's work. But since Shen Lien, a *chü-jen* from Li-yang, Chinkiang, himself died before 1855, Shen Ping-ch'eng's work would have to have been of an earlier date. It is possible that *Ts'an-sang chi-yao* was not actually written by Shen Ping-ch'eng, but only

reprinted by him. Wang Yü-hu, *Chung-kuo nung-hsueh shu-lu* (Peking, 1957), p. 195. It is also possible that he was confused with Shen Ping-chen (1679–1738), also a native of Kuei-an, perhaps his ancestor, who had authored the *Ts'an-sang yueh-fu*. See Hummel, pp. 644–645.

14. *KTSSCP*, preface 1b–2a.

15. It was supposedly based on a work by Wu Kung-chang, *Ts'an-sang chieh-hsiao shu*, which Huang admired. *TSCMCS*, 1a–b.

16. *An-chi HC* (1874), 10:46.

17. *TSYL*, 2–4. See also Wang Yü-hu, p. 200.

18. Ch'en Tso-lin, 3:2–4, in *SKYSTL*, II, 427.

19. See Chapter 4. Yen Chung-p'ing, "Ming-Ch'ing," p. 24; and *SKYSTL*, II, 15.

20. *SK*, pp. 22–24; and *Nung-hsueh pao* 6:1–6.

21. *Nung-hsueh pao* 6:5a–6a.

22. *Ibid.*, 1:7a.

23. *Shina seisan jigyō tōkeihyō*, Chiang-su, pp. 704–706.

24. *Hsin-mao shih-hsing chi* (1897), 1:39–41, in Ch'ü Chih-sheng, *Chung-kuo ku nung-shu chien-chieh* (Taipei, 1960), pp. 90–92.

25. Yen Chung-p'ing, "Ming-Ch'ing," p. 26.

26. Robert Hart, *The I. G. in Peking: Letters of Robert Hart, Chinese Maritime Customs, 1868–1907*, ed. John King Fairbank, et al. (Cambridge, Mass., 1975), p. 787.

27. *ST*, p. 76.

28. *Kōsoshō Mushakuken*, pp. 10–11, 43, says 240,000 mou out of 1,258,000. See also "Sericulture in Wusih," *Chinese Economic Journal* 1.2:142 (February 1927).

29. D. K. Lieu, pp. 59–60; Amano Motonosuke, *Shina nōson zakki* (Tokyo, 1942), pp. 99–100; *ST*, p. 77.

30. *ST*, p. 95. Compare with Yueh Ssu-ping, p. 72, which says that a mou of paddy land would cost only 30 to 40 yuan, while a mou of high-grade mulberry land would cost 150 yuan. These costs probably included the trees already planted.

31. *SK*, pp. 82–83; Yueh Ssu-ping, pp. 71–72; Tseng T'ung-ch'un, pp. 20–21.

32. *Nan-hsun CC* (1922), 31:28b.

33. *Te-ch'ing CC* (1923), 4:2–3. See also D. K. Lieu, p. 15, on similar conditions in Wu-hsing.

34. *Survey*, p. 76.

35. Murphey, *Shanghai*, p. 110.

36. *DR, 1902–1911*, I, 18, and II, 403.

37. *ST*, pp. 37–38.

38. *Chia-hsing HC* (1906), 16:6b.

39. *Shina seisan jigyō tōkeihyō*, Che-chiang, p. 1067.

40. *ST*, p. 756.

41. *TSYL*, 4a–b; *KTSSCP*, 2:47b; *HTS*, pp. 19–20; and Ch'en Heng-li, p. 188.

42. *Hu-chou FC*, quoted in Ch'en Heng-li, p. 182.

43. For example, in *Chia-hsing FC* (1906), 15:10a, and in *TSYL*, 11a.

44. *Ch'ang-hsing HC* (1892), 8:40a.

45. Ch'en Heng-li, pp. 48–49.

46. *TSYL*, 11a; *HTS*, p. 20; *TSCMCS*, 40a–b; *KTSSCP*, 2:47a–b; and *Shuang-lin CC* (1917), 14:15b; all have examples of fluctuating prices.

47. *HTS*, p. 32; and *Nan-hsun CC* (1922), 31:5b.

48. *Ch'ang-hsing HC* (1892), 8:15a–b.

49. *Survey*, p. 28; D. K. Lieu, p. 7.

50. *HTS*, p. 20; *Hu-chou FC* (1874), 30:17b; and *Wu-ch'eng HC* (1881), 28:9b.

51. *Wu-ch'ing CC* (1936), 7:17b.

52. *Survey*, p. 38.

53. *ST*, p. 84.

54. Ibid., pp. 86, 90.

55. *Chia-hsing FC* (1879), 32:22b.

56. *HTS*, pp. 19, 32; *Nan-hsun CC* (1922), 30:20a, 31:15a.

57. *Nan-hsun CC* (1922), 30:20a; *SYC: Chiang-su*, V, 160.

58. *HTS*, p. 32.

59. *Survey*, p. 39; *SZ*, XIII, 495–496.

60. *HTS*, p. 32; *Shuang-lin CC* (1917), 14:5a–b.

61. Yueh Ssu-ping, p. 22; *SYC: Chiang-su*, V, 160. *ST*, p. 85, gives details of practices at Yü-hang and other places.

62. *Survey*, p. 39.

63. *Shuang-lin CC* (1917), 14:5a.

64. *Nan-hsun CC* (1922), 31:5a.

65. *HTS*, p. 32. *Te-ch'ing HC* (1923), 4:15a, stresses the gentleman's agreement involved in these contracts.

66. *KTSSCP*, 2:47b.

67. *Shuang-lin CC* (1917), 14:15b.

68. *Survey*, p. 39.

69. *ST*, p. 85.

70. *Shuang-lin CC* (1917), 14:5b, from *Hu Ch'eng-mou FC* (Hu-chou).

71. *Survey*, p. 28.

72. *Chia-hsing FC* (1879), 32:22b; *Nan-hsun CC* (1922), 30:20a.

73. *I-wen lu*, 5/1890, in *NYSTL*, I, 428.

74. Mao Tun, pp. 29–34.

75. *KTSSCP*, 2:48b.

76. Tseng T'ung-ch'un, p. 43; *SK*, pp. 280–281; *ST*, p. 160; *Survey*, p. 91.

77. *Te-ch'ing HC* (1923), 4:2b.

78. John Lossing Buck, *Land Utilization in China* (Chicago, 1937), p. 302.

79. *ST*, pp. 82–83.

80. For examples, see *ST*, pp. 157–160; and *SK*, pp. 280–281.

81. *HTS*, p. 66; *ST*, pp. 383–386; *SKYSTL*, III, 694–695; *Ch'ang-hsing HC* (1892), 8:32a.

82. *Survey*, pp. 6–7; Yueh Ssu-ping, p. 28.

83. *HTS*, p. 2; Yueh Ssu-ping, p. 28; *SKYSTL*, II, 85.

84. D. K. Lieu, pp. 7, 59–60. Lieu's survey, however, was conducted in the early 1930s, when sericulture was already in a decline.

85. This was roughly the scale at Wusih. *Nung-hsueh pao*, 6:9.

86. Yueh Ssu-ping, pp. 160–170.

87. For example, see *Chu-lin pa-yü chih* (1932), 3:8a; and *HTS*, p. 68.

88. *HTS*, p. 68.

89. *Kōsoshō Mushakuken*, pp. 104–105, 118–119.

90. Howard and Buswell, p. 48. See also Ch. 4, n. 62.

91. Ibid., p. 168.

92. Ibid., p. 164. Increase of rice imports into Kwangtung in this period were sharp: 767,325 piculs in 1920, 8,109,725 piculs in 1921, and 13,123,937 piculs in 1922.

93. Ibid., pp. 95–105.

94. Ibid., p. 81. Also on p. 3, the authors state, "No industry in China is so subject to fluctuations in the market as sericulture."

95. For diagrams showing local variations in commercial patterns, see *Chū-Shina juyō kokubō shigen seishi chōsa hōkoku*, comp. Kō Ain Kachū renrakubu (1941), pp. 1231–1232.

96. The centers of the sapling trade were located in Shih-men hsien and Hai-ning *chou*. The latter was more important in modern times. See *SZ*, XIII, 490–491. A network of sapling hongs distributed *Hu-sang* to places as distant as Shantung, Honan, Hopei, and Hupei. Yueh Ssu-ping, pp. 21, 172.

97. *Survey*, p. 57.

98. *ST*, p. 98.

99. *Chu-lin pa-yü chih* (1932), 3:7a. According to this source, Yü-hang eggs had a better yield, but were more susceptible to disease. *SYC: Che-chiang*, IV, 192, also lists Hsin-ch'ang, Fu-yang, Wu-hsing, and other places as egg centers. There are tables on pp. 192–212 about egg commerce.

100. *Survey*, pp. 54–60; and *ST*, pp. 102–107.

101. *ST*, p. 114.

102. *SYC: Che-chiang*, IV, 202.

103. *Survey*, p. 56; *ST*, p. 111.

104. See Ch. 1. Also, *Chen-tse CC* (1844), 2:1b–2a; *Nan-hsun CC* (1922), 32:20a–21b.

105. Nan-hsun CC (1922), 33:3b.

106. *Shuang-lin CC* (1917), 17:1a–b.

107. *Ch'ang-hsing HC* (1892), 8:37a.

108. *Nan-hsun CC* (1922), 30:21a.

109. *ST*, p. 387. *Survey*, p. 72, reported that Chen-tse had 20 hongs for the export trade.

110. *Wu-ch'ing CC* (1936), 21:7b–8b.

111. *Nan-hsun CC* (1922), 19b–20a; *HTS*, pp. 79–80; *Shuang-lin CC* (1917), 16:6b. These sources give the names of other types of hongs.

112. Horie Eiichi, *Shina sanshigyō ni okeru torihiki kankō* (Tokyo, 1944), pp. 129–131. Also, *SZ*, XIII, 600–604.

113. *Survey*, p. 71; *SZ*, XIII, 600–601; and *SYC: Che-chiang*, VII, 46.

114. *SZ*, XV, 735–736. There is a list of Nan-hsun hongs here.

115. *ST*, p. 388. *SZ*, XV, 735, gives 30–40 as the total number of *ching-ssu-hang* in Nan-hsun or Chen-tse.

116. *ST*, pp. 386–387.

117. *SZ*, XV, 735–736, and also XIII, 607–608, for details about the warehouses.

118. *SZ*, XV, 729; *ST*, pp. 395–396; Negishi Tadashi, pp. 249–250.

119. *Shuang-lin CC* (1917), 15:3a–b.

120. *HTS*, p. 26, and also *Chen-tse CC* (Tao-kuang ed.), 2:10, in *SKYSTL*, I, 478.

121. *Ch'ang-hsing HC* (1892), 8:13a, 8:37b.

122. See descriptions in *KTSSCP*, original preface 1b–2a; *HTS*, p. 80; *Nan-hsun CC* (1922), 31:27b.

123. *Wu-hsing nung-ts'un ching-chi*, pp. 123–124, in *SKYSTL*, II, 83–84.

124. Remer, p. 133.

SIX   *Foreign Trade and Modern Enterprise*

1. *ATRR 1885*, p. 163.

2. Feuerwerker, *The Chinese Economy, ca. 1870–1911*, pp. 32, 38.

3. British Consular report, Shanghai, 1872, in *KYSTL*, IA, 68; and Shen Wen-wei, *Chung-kuo ts'an-ssu-yeh she-hui yü she-hui hua ching-ying*, 1/1937, in *KYSTL*, IVA, 109. Other sources put the opening date of this filature in 1859. See *KYSTL*, IA, 65.

4. Yueh Ssu-ping, pp. 5–6; Shen Wen-wei in *KYSTL*, IVA, 110; Howard and Buswell, p. 6; H. D. Fong, p. 492.

5. *ST*, pp. 944–945. Of the 202, 154 were in Shun-te, 45 in Nan-hai, and only 3 were elsewhere. H. D. Fong, p. 492, gives different figures. Discrepancies can easily be accounted for by the fact that not all filatures operated every year, and the rate of failure was quite high.

6. Yueh Ssu-ping, pp. 242–251, gives a list of Canton filatures. Howard and Buswell, pp. 121–122; H. D. Fong, pp. 492–493; Feuerwerker, *The Chinese Economy, ca. 1870–1911*, pp. 38, 41.

7. *KYSTL*, IVA, 65.

8. *Chieh-pao*, 8/1882, in *KYSTL*, IA, 70. Nihon tōa dōbunkai, *Kōnan jijō*, in *KYSTL*, IA, 73.

9. Shih Min-hsiung, pp. 83–84; *KYSTL*, IA, 65; *Chieh-pao*, 7/1902, in *KYSTL*, IB, 971; H. D. Fong, p. 493.

10. Nihon tōa dōbunkai, 1910, in *KYSTL*, IA, 73.

11. *DR, 1892–1901*, I, 511, in H. D. Fong, p. 494.

12. See Table 24. Yang Ta-chin, I, 126–132, has a list of 104 filatures in Shanghai around 1931, which gives the name of each factory, the date of establishment, capitalization, number of workers, reels, cocoons used, and amount of silk produced.

13. *ST*, p. 236.

14. *SK*, p. 134.

15. H. D. Fong, p. 496. Yang Ta-chin, I, 132, states that the Yü-ch'ang was established in 1901. *Chung-hua jih-pao*, 6/1941, in *KYSTL*, IVA, 175–178, states that Wusih's first filature was built in 1905.

16. H. D. Fong, p. 496.

17. *ST*, pp. 355–356. *Survey*, p. 33, has a comparison of Shanghai and Wusih wages. For a complete list and data on Wusih filatures, see Yang Ta-chin, I, 132–136.

18. *ST*, pp. 352–354.

19. *ST*, p. 360.

20. *Ts'an-ssu tsa-chih*, 1948, in *KYSTL*, IVA, 181.

21. *Te-ch'ing HC* (1923), 13:15b.

22. *Shih-wu pao*, 10/1896, in *KYSTL*, IIB, 695. *Chung-wai jih-pao*, 5/1900, in *KYSTL*, IIA, 406, reports the failure of these attempts.

23. *Ts'an-ssu tsa-chih*, 1948, in *KYSTL*, IVA, 180.

24. *Survey*, p. 64. Hangchow modern enterprises were particularly subject to Japanese influence; most of them had Japanese instructors as well as Japanese equipment. *ST*, p. 364.

25. *Chia-hsing hsien nung-ts'un tiao-ch'a*, pp. 77–78.

26. H. D. Fong, pp. 497–498. Yang Ta-chin, I, 136–138, has a complete list of Chekiang filatures and relevant data.

27. *ST*, pp. 361–362; Yueh Ssu-ping, pp. 121–123; *SZ*, XV, 752–753.

28. *ST*, p. 239. The total number of filatures in Kiangnan in 1928 was 160 according to this survey, but on p. 234 the subtotals add up to 162.

29. *Shih-wu pao*, 4/1897, in *KYSTL*, IIB, 692.

30. Feuerwerker, *The Chinese Economy, ca. 1870-1911*, p. 41. One tael was roughly 1.50 yuan. See exact rates in Tuan-liu Yang and Hou-pei Hou, p. 151.

31. Shih Min-hsiung, pp. 85-86, from *KZ*, XII, 35.

32. Yang Ta-chin, I, 112.

33. *SYC: Chiang-su*, V, 167-182; *SYC: Che-chiang*, VII, 40-43; *ST*, pp. 231-232.

34. Li Hung-chang letters, 9/1883, in *KYSTL*, IA, 72.

35. Li Hung-chang naval letters, 1/1887, in ibid.

36. *Chang Wen-hsiang kung ch'üan-chi*, in *KYSTL*, IIA, 592-593, 595; *DR, 1892-1901*, p. 545.

37. Yueh Ssu-ping, pp. 121-123; *SZ*, XV, 752-753.

38. Marianne Bastid, "Le développement des filatures de soie modernes dans la province du Guangdong avant 1894," *The Polity and Economy of China: The Late Professor Yuji Muramatsu Commemoration Volume* (Tokyo, 1975), pp. 175-188.

39. *ST*, pp. 235-236.

40. *Yuan-tung ching-chi fa-chan chung ti wai-kuo ch'i-yeh*, in *KYSTL*, IIA, 279-280.

41. Tientsin *Ta-kung pao*, 4/1917, in *KYSTL*, IVA, 171; *SK*, p. 130.

42. *ST*, p. 333.

43. Yueh Ssu-ping, pp. 131-137, passim; and *SZ*, XV, 745-752.

44. Tseng T'ung-ch'un, pp. 164-166.

45. *Ying-hang chou pao*, 6/1929, in *NYSTL*, III, 453.

46. Shih Min-hsiung, p. 97.

47. D. K. Lieu, p. 101.

48. Yueh Ssu-ping, p. 92; and Tseng T'ung-ch'un, p. 60.

49. Ibid., pp. 38-39, 92. According to the 1925 survey of the Shanghai International Testing House, the average rental cost of one basin per year was 33 taels, while the average capital investment per basin was 291 taels. *Survey*, p. 19. *SYC: Chiang-su*, VIII, 116, states that 260 taels were needed.

50. *Chung-kuo ts'an-ssu ch'an-hsiao ch'ing-k'uang ti tiao-ch'a*, in *KYSTL*, IVA, 102. Horie Eiichi, p. 104, estimated that about 70% of Wusih's filatures were rented out.

51. *Survey*, p. 34.

52. D. K. Lieu, pp. 98-99. See also Yang Ta-chin, I, 114; *SYC: Chiang-su*, V, 167-182; and *SYC: Che-chiang*, VII, 40-42.

53. *ST*, p. 321.

54. *ST*, pp. 338-340; Yueh Ssu-ping, p. 202.

55. *SZ*, XV, 687-688; Tientsin *Ta-kung pao*, 4/1917, in *KYSTL*, IVA, 171.

56. *Ch'i-yeh chou-k'an,* 11 and 12/1943, in *KYSTL,* IVA, 113.

57. *Survey,* p. 82.

58. D. K. Lieu, p. 91.

59. *ST,* p. 237.

60. *DR, 1922–1931,* II, 25–26, quoted in H. D. Fong, pp. 495–496.

61. *ST,* p. 990.

62. Howard and Buswell, pp. 121, 130.

63. Duran, p. 147.

64. Marjorie Topley, "Marriage Resistance in Kwangtung," in Margery Wolf and Roxane Witke, eds., *Women in Chinese Society* (Stanford, 1975), pp. 67–88. Also, Robert Y. Eng, "Silk Reeling Enterprises in Canton and Shanghai," paper for the Conference on the Colonial Port City in Asia (June 1976), p. 15.

65. Yueh Ssu-ping, pp. 107–109.

66. Ibid., p. 101.

67. D.K. Lieu, pp. 118–120. See also *ST,* p. 252; Yang Ta-chin, I, 126–144.

68. D. K. Lieu, pp. 124–131; *Survey,* p. 18.

69. D. K. Lieu, pp. 131–134; Yueh Ssu-ping, pp. 101–104; *SZ,* XV, 702–703.

70. Yueh Ssu-ping, p. 96, gives 10–15 as the average age.

71. Eleanor M. Hinder, *Life and Labour in Shanghai: A Decade of Labour and Social Administration in the International Settlement* (New York, 1944), p. 35.

72. Tientsin *Ta-kung pao,* 4/1917, in *KYSTL,* IVA, 172.

73. Hinder, p. 36. For more on the conditions of workers, see *ST,* pp. 299–319. For more information about the costs of operating a filature, see *SZ,* XV, 689–690; Yueh Ssu-ping, pp. 91–100, 111–113; *ST,* pp. 341–345.

74. D. K. Lieu, pp. 224–225.

75. Jean Chesneaux, *The Chinese Labor Movement, 1919–1927* (Stanford, 1968), pp. 57–61.

76. *Survey,* pp. 2, 19.

77. *ST,* p. 166. For a physical description of cocoon hongs, see *ST,* p. 169, and *Survey,* pp. 10–12.

78. *SZ,* XV, 533–534.

79. *ST,* pp. 162–163. Horie Eiichi, pp. 41 and 174, uses the pejorative expression, *t'u-hao lieh-sheng,* or "local bullies and evil gentry," to describe the owners.

80. Negishi Tadashi, p. 234; Horie Eiichi, p. 41; *SYC: Chiang-su,* V, 191.

81. *SYC: Chiang-su,* V, 191.

82. Horie Eiichi, p. 41.

83. Yueh Ssu-ping, pp. 32, 48.

84. Horie Eiichi, p. 43.

85. *ST,* p. 177; *SYC: Che-chiang,* IV, 213; *Survey,* p. 13.

86. Horie Eiichi, p. 46.

87. Yueh Ssu-ping, p. 83.

88. *ST,* pp. 177-178, shows a sample contract. Horie Eiichi, pp. 47-50.

89. *ST,* pp. 179-182.

90. Horie Eiichi, pp. 47-50.

91. Ibid., p. 51; *ST,* pp. 177-182.

92. Buchanan, pp. 8-9.

93. *Survey,* p. 9. According to another estimate, about 60% of the dried cocoons marketed in Chekiang and Kiangsu in the mid-1920s were purchased directly by the filatures, and 40% were handled by the brokers. Buchanan, p. 8. According to a third estimate, however, the filatures bought only 30 or 40% directly, and the rest were handled by brokers. Horie Eiichi, p. 38.

94. Ibid., and D. K. Lieu, p. 104.

95. Yueh Ssu-ping, pp. 35-37, gives cocoon export prices for 1862-1932.

96. *Survey,* p. 13.

97. Buchanan, pp. 3-7.

98. *ST,* pp. 164, 221-223; *Survey,* pp. 63-64.

99. *Survey,* p. 13.

100. At Shanghai this would result in a rise in interest rates and also in the lowering of the value of the tael with respect to the yuan, according to Buchanan, p. 7.

101. *ST,* p. 183.

102. Yueh Ssu-ping, pp. 201-202.

103. Tseng T'ung-ch'un, p. 47; Yueh Ssu-ping, p. 84.

104. Buchanan, p. 9.

105. Yueh Ssu-ping, pp. 110-111.

106. On the relationship between *ch'ien-chuang* and foreign banks, see McElderry, pp. 21-22.

107. *ST,* pp. 228-229, has a list of important warehouses associated with major banks and *ch'ien-chuang.* Yueh Ssu-ping, pp. 110-111, also has such a list.

108. Buchanan, p. 12.

109. Ibid., p. 10.

110. Yueh Ssu-ping, p. 85.

111. *ST,* p. 170.

112. *ST,* pp. 396-398; Shanghai International Testing House, *China Silk,* p. 6; *Survey,* p. 86; *NYSTL,* II, 170-171, 194.

113. *ST*, pp. 174–175.

114. See details in *ST*, pp. 175–176. There is no evidence that this policy was ever implemented on a wide scale.

115. Buchanan, pp. 4–6.

116. *ST*, p. 176.

117. Yueh Ssu-ping, p. 88; *Hang-chou shih ching-chi tiao-ch'a*, 6:17; *Te-ch'ing HC* (1923), 4:32b; *Wu-ch'ing CC* (1936), 22:9b; *SYC: Che-chiang*, IV, 213.

118. Ibid., and *Wu-ch'ing CC* (1936), 22:13a.

119. *ST*, p. 215. Compare *Survey*, p. 31.

120. *ST*, pp. 214–220, gives a detailed analysis of the costs involved in processing cocoons. These figures more or less correspond with those given in *Survey*, p. 13.

121. *ST*, p. 225, gives prices of dry cocoons at Shanghai between 1916 and 1924.

122. *ST*, p. 454.

123. Yueh Ssu-ping, pp. 41, 165–166; and *ST*, pp. 10–11.

124. *Hang-chou shih ching-chi tiao-ch'a*, 6:17.

125. *ST*, p. 453. The actual income of the Kiangsu provincial government in 1923 was about 32 million yuan. *Ko-sheng-ch'ü li-nien ts'ai-cheng hui-lan*, comp. Ts'ai-cheng pu, ts'ai-cheng tiao-ch'a ch'u (Peking, 1927), 1.1:7. The income of the Chekiang provincial government in 1924 was about 18 million yuan. Ibid., 1.2:7, 10.

126. It was observed that in the years when the price of filature silk was very high, the quantity of cocoons exported had also been very large. Tseng T'ung-ch'un, pp. 101–104. Yueh Ssu-ping, pp. 34–37, has a table of cocoon exports and prices, 1862–1922.

127. *Te-ch'ing HC* (1923), 2:14a, reported how peasants sold "all" their cocoons to the local hongs when they were established. *Nan-hsun CC* (1922), 30:21a; *Survey*, p. 74.

128. *Survey*, p. 63.

129. Buchanan, p. 4; Yueh Ssu-ping, p. 86.

130. *ATRR 1898*, pp. 242–243.

131. Banister, p. 162.

132. *DR, 1912–1921*, p. 366; *ATRR 1897*, p. 237.

133. Yueh Ssu-ping, pp. 238–239.

134. Howard and Buswell, p. 144.

135. Ibid., pp. 105–111; *ST*, p. 965.

136. Howard and Buswell, pp. 147–148. Also, *ST*, p. 1004.

137. In his dissertation, "Imperialism and the Chinese Economy," Eng considers what factors might have accounted for Canton's faster start in the silk filature business, citing relative freedom from foreign capital as a

possible cause (pp. 71–75). He points out that the money shops of Canton performed the same financing functions that the Shanghai *ch'ien-chuang* did, but they had an additional and plentiful source of capital from the remissions from local sons who had emigrated to Southeast Asia (pp. 100–101).

138. One attempt to show quantitatively that there was an economic surplus is found in Carl Riskin, "Surplus and Stagnation in Modern China," in Dwight H. Perkins, ed., *China's Modern Economy in Historical Perspective* (Stanford, 1975), pp. 49–82.

139. G. C. Allen and Audrey C. Donnithorne, *Western Enterprise in Far Eastern Development* (London, 1954), p. 68.

140. *Survey*, p. 9.

141. On Lo Chen-yü and this organization, see *Wu-hsu pien-fa*, ed. Chung-kuo shih-hsueh hui (Shanghai, 1953), IV, 427–431.

142. *Nung-hsueh pao* 1:7 (1897).

143. Hsiao Wen-hsiao memorial, 7/1898, as found in *Wu-hsu pien-fa tang-an shih-liao*, comp. Kuo-chia tang-an chü, Ming-Ch'ing tang-an kuan (Peking, 1958), pp. 397–400.

144. Memorial of 7/1898, in ibid., pp. 404–406. On Tuan Fang, see Hummel, p. 780.

145. *Nung-hsueh pao* 1:7, 21:3a–b (1897).

146. *SK*, p. 24.

147. *DR, 1902–1911*, p. 48.

148. *Survey*, pp. 52–53.

149. *Hang-chou shih ching-chi tiao-ch'a*, 6:8–12.

150. Yueh Ssu-ping, pp. 53, 58; *Survey*, pp. 40–41. This school is described by Fei Hsiao-t'ung, *Peasant Life in China: A Field Study of Country Life in the Yangtze Valley* (London, 1939), p. 205ff.

151. *ST*, pp. 500–502; *Survey*, p. 45.

152. *DR, 1922–1931*, I, 623.

153. Yueh Ssu-ping, p. 56ff.

154. *ST*, pp. 488–489; on pp. 490–491 there is a list of schools in Kiangsu and Chekiang.

155. Howard and Buswell, pp. 8–11.

156. Yueh Ssu-ping, pp. 57–58.

157. *SYC: Che-chiang*, IV, 219–220.

158. Yueh Ssu-ping, pp. 57–58; D. K. Lieu, p. xvi.

159. *SYC: Che-chiang*, IV, 222–224.

160. See Yueh Ssu-ping, pp. 24, 187–190. *SYC: Che-chiang*, IV, 207, has a list of centers.

161. *Chia-hsing hsien nung-ts'un tiao-ch'a*, pp. 72–73.

162. D. K. Lieu, p. 32.

163. Hummel, I, 331.

164. *ST*, pp. 325–326.

165. *Mushaku no seishigyō*, pp. 3–7. Hsueh Shou-hsuan was related by marriage to the Jung family, Wusih's flour-mill magnates. *Chung-hua jih-pao*, 6/1941, in *KYSTL*, IVA, 176–177.

166. *NYSTL*, III, 164–166, 487–488.

167. *NYSTL*, II, 490–492; Horie Eiichi, p. 68.

168. *SYC: Che-chiang*, IV, 227.

169. Horie Eiichi, pp. 23–24.

170. Ibid., pp. 63–73.

171. Fei Hsiao-t'ung, pp. 218–228.

172. *Survey*, pp. 85–86; Yueh Ssu-ping, p. 56; Yin Liang-ying, *Chung-kuo ts'an-yeh shih* (Nanking, 1931), p. 14.

173. Yueh Ssu-ping, p. 57; See *ST*, pp. 514–532, for details about the activities of this committee.

174. Yueh Ssu-ping, p. 59.

175. Buchanan, pp. 40–41.

176. Ibid., pp. 31–33.

177. Ibid., pp. 40–41. See *ST*, pp. 533–539, for details on the Shanghai Testing House.

178. Yueh Ssu-ping, p. 57; D. K. Lieu, p. xvi.

179. Li Shan-shu, *K'ao-ch'a Ou-Mei sheng-ssu shih-ch'ang pao-kao* (Soochow, 1937), p. 73.

180. *Ssu-yeh chih yu* 28:170–173 (1940).

Conclusion

1. Table 9, and Liang-lin Hsiao, pp. 22–24, 38–39.

2. Most of the following material on Japan is taken from "The Silk Export Trade and the Economic Modernization of China and Japan," a paper originally prepared for the Conference on American-East Asian Economic Relations, Mt. Kisco, N.Y., June 1976. A revised version will be published under the title "Silks by Sea: Trade, Technology, and Enterprise in China and Japan," in the winter 1981 issue of *The Business History Review*. I am grateful to Yoshida Kazuko, a graduate student in the Department of International Relations at Tokyo University, for assistance in research for the revised version. Ms. Yoshida's undergraduate thesis on the Japanese silk industry has been published as an article in Japan. See note 18 below. I also wish to thank Professor Richard J. Smethurst, Department of History, University of Pittsburgh, for insightful comments on this chapter.

3. Keishi Ohara, p. 262; Kenzō Hemmi, p. 316; and Yamaguchi Kazuo, ed., *Nihon sangyōkin'yūshi kenkyū: seishi kin'yūhen* (Tokyo, 1966), pp. 514–623 passim.

4. Yagi Haruo, *Nihon keizaishi gaisetsu* (Tokyo, 1974), p. 260; and Kenzō Hemmi, p. 316.

5. Keishi Ohara, p. 263; and Hugh T. Patrick, "Japan, 1868–1914," in Rondo E. Cameron, ed., *Banking in the Early Stages of Industrialization* (New York, 1967), p. 279.

6. Patrick, p. 249, feels that government leadership in developing the banking system has been exaggerated; he prefers to stress the importance of private initiative.

7. Miyohei Shinohara, "Economic Development and Foreign Trade in Pre-War Japan," in C. D. Cowan, ed., *The Economic Development of China and Japan* (New York and London, 1964), p. 226.

8. See map in Huber, end fold-out.

9. Lockwood, p. 45.

10. *Taikei Nihonshi sōsho,* comp. Yamakawa shuppansha (Tokyo, 1965), XI, 91, 278. Also Yagi Haruo, "Seishigyō," I, 223. Also, Fujimoto Jitsuya, I, 147–179 passim.

11. The sericultural manuals in the early Tokugawa period were largely based on Chinese works, but those of the mid-Tokugawa period were usually based on Japanese innovations. Uchida Hoshimi, *Nihon bōshoku gijitsu no rekishi* (Tokyo, 1960), pp. 54–55.

12. Yagi Haruo, "Seishigyō," I, 236. The names used here are those of current prefectures. Traditionally Ōshu corresponded roughly to the Tōhōku region including Fukushima, Joshu to Gumma, and Shinshu to Nagano.

13. A summer crop was developed in Nagano in the 1830s. *Nihon sangyōshi taikei,* V, 186. For a discussion of other Tokugawa technological advances, see Yagi Haruo, "Seishigyō," I, 240; and Uchida Hoshimi, pp. 57–66.

14. *Nihon sangyōshi taikei,* IV, 260–261; and Yagi Haruo, "Seishigyō," I, 227.

15. For a description in English of such activity in Saitama prefecture, see William Jones Chambliss, *Chiaraijima Village: Land Tenure, Taxation, and Local Trade, 1818–1884* (Tucson, Arizona, 1965), pp. 17–22.

16. Keishi Ohara, pp. 229–233; and Thomas C. Smith, *Political Change and Industrial Development in Japan: Government Enterprise, 1868–1880* (Stanford, 1955), pp. 58–61.

17. See for example, *Taikei Nihonshi sōsho,* XII, 226–229.

18. Yoshida Kazuko, "Meiji shoki no seishi gijitsu ni okeru dochaku to gairai," *Kagakushi kenkyū,* 2.16:19 (Spring 1977). Regional differences are also explored in Ebato Akira, *Sanshigyō chiiki no keizai chirigakuteki kenkyū* (Tokyo, 1969).

19. Keishi Ohara, pp. 243–249; Yagi Haruo, "Seishigyō," I, 240, notes that the Japanese technique was yet unknown in Europe or China. In Kiangnan, rereeled silk did not become an important item until the 1910s.

20. *Nihon sen'i sangyōshi,* comp. Nihon sen'i kyōgikai (Osaka, 1958), I, 942. If this data did not include waste silk, the proportion of machine-reeled silk would be close to 90%. See also *Nihon boeki seiran,* comp. Tōyō keizai shimpōsha, reprint (Tokyo, 1975), pp. 53–55, which shows that virtually all exports in 1925 were steam filatures.

21. *Taikei Nihonshi sōsho,* XII, 399–401.

22. Ishii Kanji, *Nihon sanshigyōshi bunseki* (Tokyo, 1972), p. 88; and Keishi Ohara, pp. 275–276.

23. Huber, p. 42.

24. Johannes Hirschmeier, *The Origins of Entrepreneurship in Meiji Japan* (Cambridge, Mass., 1964), p. 95, makes this point, as do others. The dualism which is said to have characterized the Japanese economy in modern times was not, in my opinion, found in the silk industry.

25. For example in Yamanashi. See Nakamura Masanori, "Seishigyō no tenkai to jinushisei," *Shakai keizai shigaku,* 32:46–71 (1967). In Gumma and other places ex-samurai were leading entrepreneurs. See *Nihon sangyōshi taikei,* IV, 277–278.

26. Yagi Haruo, *Nihon keizaishi gaisetsu,* p. 217; and also Yamaguchi Kazuo, pp. 10–11.

27. Keishi Ohara, pp. 292–294. Also, Huber, p. 17.

28. Ishii Kanji, p. 422ff.

29. This is the principal argument of Frances V. Moulder, *Japan, China, and the modern world economy* (Cambridge, England, 1977).

30. It is significant that in Moulder's entire discussion of foreign trade as a form of "incorporation" (pp. 98–110), the Chinese silk export trade is not mentioned. In her discussion of foreign investment in Chinese manufacturing, however, Moulder does admit that Western firms were unsuccessful in the silk reeling industry (p. 113).

Bibliography

Allen, G. C. and Audrey C. Donnithorne. *Western Enterprise in Far Eastern Development.* London, George Allen and Unwin, 1954.

Amano Motonosuke 天野元之著 . *Shina nōson zakki* 支那農村襍記 (Miscellaneous notes on Chinese rural villages). Tokyo, 1942.

An-chi hsien chih 安吉縣志 (Gazetteer of An-chi hsien). 1874.

Annual Report. Silk Association of America. New York, 1873-1930.

Annual Trade Report and Returns, comp. China, Maritime Customs. Shanghai, Inspectorate General of Customs, 1868-1930.

Atwell, William S. "Notes on Silver, Foreign Trade, and the Late Ming Economy," *Ch'ing-shih wen-t'i* 3.8:1-33 (December 1977).

Banister, T. R. "A History of the External Trade of China, 1834-1881," in China, Maritime Customs, comp., *Decennial Reports, 1922-1931.* 2 vols. Shanghai, 1933, I, 1-193.

Bastid, Marianne. "Le développement des filatures de soie modernes dans la province du Guangdong avant 1894," in *The Polity and Economy of China: The Late Professor Yuji Muramatsu Commemoration Volume.* Tokyo, Tōyō keizai shinposha, 1975.

Bergère, Marie-Claire. *Une crise financière à Shanghai à la fin de l'ancien régime.* Paris, Mouton, 1964.

Boxer, C. R. *The Great Ship from Amacon: Annals of Macao and the Old Japan Trade, 1555-1640.* Lisbon, Centro de Estudos Historicos, Ultramarinos, 1963.

Buchanan, Ralph E. *The Shanghai Raw Silk Market.* New York, 1929.

Buck, John Lossing. *Land Utilization in China.* Chicago, University of Chicago Press, 1937.

"Canton Exports of Raw Silk and Silk Waste in 1927," *Chinese Economic Journal* 2.6:529–532 (June 1928).

Chambliss, William Jones. *Chiaraijima Village: Land Tenure, Taxation, and Local Trade, 1818–1884.* Tucson, University of Arizona Press, 1965.

Chang K'ai 章楷 . *Ts'an-yeh shih-hua* 蠶業史話 (Tales about sericulture). Peking, Chung-hua shu-chü, 1979.

Chang Kwang-chih. *The Archaeology of Ancient China.* 3rd ed. New Haven, Yale University Press, 1977.

Ch'ang-hsing hsien chih 長興縣志 (Gazetteer of Ch'ang-hsing hsien). 1892.

Chao Kang. *The Development of Cotton Textile Production in China.* Cambridge, Harvard University, East Asian Research Center, 1977.

Chao Ya-shu 趙雅書 . "Sung-tai ts'an-ssu-yeh ti ti-li fen-pu," 宋代蠶絲業的地理分佈 (The geographical distribution of the silk industry in the Sung dynasty), *Shih-yuan* 史原 (National Taiwan University) 3:65–94 (September 10, 1972).

Che-chiang sheng nung-ts'un tiao-ch'a 浙江省農村調查 (A survey of rural Chekiang), ed. Chung-kuo hsing-cheng yuan, nung-ts'un fu-hsing wei-yuan hui 中國行政院農村復興委員會 (Committee on Rural Reconstruction, Legislative Yuan, China). Shanghai, 1934.

Chen-tse chen chih 震澤鎮志 (Gazetteer of Chen-tse chen). 1844.

Ch'en Heng-li 陳恆力 . *Pu Nung-shu yen-chiu* 補農書研究 (Research on the *Pu Nung-shu*). Peking, Chung-hua shu-chü, 1958.

Ch'en Shih-ch'i 陳詩啟 . *Ming-tai kuan shou-kung-yeh ti yen-chiu* 明代官手工業的研究 (A study of the official handicraft industry of the Ming dynasty). Wuhan, Hu-pei jen-min ch'u-pan she, 1958.

Ch'en Tso-lin 陳作霖 . "Feng-lu hsiao-chih" 鳳麓小志 (Record of Feng-lu) in his *Chin-ling so-chih wu-chung* 金陵瑣志五種 (Five fragmentary records of Nanking). 9 chüan. 1900.

Chesneaux, Jean. *The Chinese Labor Movement, 1919–1927.* Translated from the French by H. M. Wright. Stanford, Stanford University Press, 1968.

Chi-min yao-shu chin-shih 齊民要術今釋 (A modern translation of the *Chi-min yao-shu*), ed. and trans. Shih Sheng-han 石聲漢. 4 vols. Peking, K'o-hsueh ch'u-pan she, 1957.

Chia-hsing fu chih 嘉興府志 (Gazetteer of Chia-hsing prefecture). 1879.

Chia-hsing hsien chih 嘉興縣志 (Gazetteer of Chia-hsing hsien). 1906.

Chia-hsing hsien nung-ts'un tiao-ch'a 嘉興縣農村調查 (Investigation of rural villages in Chia-hsing hsien), ed. Feng Tz'u-kang 馮紫崗 . Hangchow, National Chekiang University, 1936.

Chiang-su sheng Ming-Ch'ing i-lai pei-k'o tzu-liao hsuan-chi 江蘇省 明清以來碑刻資料選集 (A selection of stone inscription materials of Kiangsu province since the Ming and the Ch'ing), comp. Chiang-su po-wu kuan 江蘇博物館 (Museum of Kiangsu) Peking, 1959. Reprint, Tokyo, Daian, 1967.

Chiang-su sheng shih-yeh hsing-cheng pao-kao shu 江蘇省實業行 政報告書 (Report on the administration of industrial enterprises in Kiangsu province), comp. Chiang-su sheng hsing-cheng kung-shu shih-yeh-ssu 江蘇省行政公署實業司 (The industrial section of the administration department of Kiangsu province). 1914.

Ch'ing chiu-ch'ao ching-sheng pao-hsiao ts'e mu-lu 清九朝京省報 銷冊目錄 (Catalog of accounts forwarded to the capital during the nine reigns of the Ch'ing dynasty). Peking, 1935.

Ch'ing shih-lu ching-chi tzu-liao chi-yao 清實錄經濟資料 集要 (A compendium of economic materials in the *Veritable Records* of the Ch'ing dynasty), ed. Nan-k'ai ta-hsueh, li-shih hsi 南開大學 歷史系 (History Department, Nankai University). Peking, Chung-hua shu-chü, 1959.

Chu-chi hsien chih 諸暨縣志 (Gazetteer of Chu-chi hsien). 1903, 1911.

Chu Ch'i-ch'ien 朱啟鈐 . *Ssu-hsiu pi-chi* 絲繡筆記 (Notes on silk embroidery). 2 chüan. 1930.

Chu-lin pa-yü chih 竹林八圩志 (Gazetteer of Chia-hsing hsien). 1932.

Chū-Shina juyō kokubō shigen seishi chōsa hōkoku 中支那重要國 防資源生絲調查報告 (Report on the investigation of raw silk in central China as an important national defense resource). Kō Ain Kachū renrakubu 興亞院華中連絡部 (Central China Bureau of Asia Development Office). 1941.

Ch'ü Chih-sheng 曲直生 . *Chung-kuo ku nung-shu chien-chieh* 中 國古農書簡介 (Simplified explanation of ancient Chinese agricultural manuals). Taipei, Ching-chi yen-chiu she, 1960.

Ch'üan Han-sheng 全漢昇 . "Mei-chou pai-yin yü shih-pa shih-chi,

Chung-kuo wu-chia ko-ming ti kuan-hsi" 美州白銀與十八世紀中國物價革命的關係 (American silver and the eighteenth-century price revolution in China), *Chung-yang yen-chiu yuan li-shih yü-yen yen-chiu so ch'i-k'an* 中央研究院歷史語言研究所集刊 (Bulletin of the Institute of History and Philology of the Academia Sinica) 28:517–550 (May 1957).

———. "Sung-Ming chien pai-yin kou-mai-li ti pien-tung chi ch'i yuan-yin" 宋明間白銀購買力的變動及其原因 (Fluctuations in the purchasing power of silver from the Sung to the Ming and their causes), *Hsin-ya hsueh-pao* 新亞學報 (New Asia journal) 8.1:157–186 (February 1967).

———. "Tzu Ming-chi chih Ch'ing chung-yeh Hsi-shu Mei-chou ti Chung-kuo ssu-huo mao-i" 自明季至清中葉西屬美州的中國絲貨貿易 (China's silk trade with Spanish America from the Ming to the mid-Ch'ing period) in his *Chung-kuo ching-chi shih yen-chiu* 中國經濟史研究 (Researches on Chinese economic history). 3 vols. Hong Kong, Hsin-ya yen-chiu so, 1976. I, 451–473.

Chung-kuo chin-tai kung-yeh shih tzu-liao 中國近代工業史資料 (Materials on the history of industry in modern China). 4 series, 2 volumes each. 1st series, ed. Sun Yü-t'ang 孫毓棠 ; 2nd series, ed. Wang Ching-yü 汪敬虞 ; 3rd and 4th series, ed. Ch'en Chen 陳真 . Peking, San-lien shu-tien, 1957–1961.

Chung-kuo chin-tai kuo-min ching-chi shih chiang-i 中國近代國民經濟史講義 (Lectures on the economic history of the modern Chinese people), ed. Hu-pei ta-hsueh cheng-chih ching-chi-hsueh chiao-yen shih 湖北大學政治經濟學教研室 (Department of Politics and Economics, Hupei University). Peking, Kao-teng chiao-yü ch'u-pan she, 1958.

Chung-kuo chin-tai nung-yeh shih tzu-liao 中國近代農業史資料 (Materials on the agricultural history of modern China), ed. Li Wen-chih 李文治 . 3 vols. Peking, San-lien shu-tien, 1957.

Chung-kuo chin-tai shou-kung-yeh shih tzu-liao 中國近代手工業史資料 (Materials on the history of the handicraft industry in modern China), ed. P'eng Tse-i 彭澤益 . 4 vols. Peking, San-lien shu-tien, 1957.

Chung-kuo shih-yeh chih: Che-chiang sheng 中國實業誌:浙江省 (Gazetteer of Chinese indstury: Chekiang province), comp. Chung-kuo shih-yeh pu, kuo-chi mao-i chü 中國實業部，國際貿易

局　(Chinese Ministry of Industry, Bureau of International Trade). Shanghai, 1933.

Chung-kuo shih-yeh chih: Chiang-su sheng 中國實業誌：江蘇 省 (Gazetteer of Chinese industry: Kiangsu province), comp. Chung-kuo shih-yeh pu, kuo-chi mao-i chü 中國實業部國際貿易 局　(Chinese Ministry of Industry, Bureau of International Trade). Shanghai, 1933.

Chung-kuo tzu-pen chu-i meng-ya wen-t'i t'ao-lun chi 中國資本主 義萌芽問題討論集　(Collection of debates on the question of the "sprouts of capitalism" in China), comp. Chung-kuo jen-min ta-hsueh, Chung-kuo li-shih chiao-yen shih 中國人民大學 中國歷史教研室　(Chinese People's University, Chinese History Department), 2 vols. Peking, San-lien shu-tien, 1957.

Cooper, Michael. "The Mechanics of the Macao-Nagasaki Silk Trade," *Monumenta Nipponica* 27.4:423–433 (Winter 1972).

Decennial Reports, comp. China, Maritime Customs. Shanghai, Inspectorate General of Customs, 1882–1931.

Dermigny, L. *La Chine et l'occident: le commerce à Canton au XVIIIe siècle, 1719–1833.* 3 vols. Paris, S.E.V.P.E.N., 1964.

Dobb, Maurice. *Studies in the Development of Capitalism.* Rev. ed. New York, International Publishers, 1963.

Du Halde, Jean Baptiste. *The General History of China.* Translation of *Description géographique, historique, politique, et physique de l'empire de la Chine et de la Tartarie chinoise.* 4 vols. Paris, Lemercier, 1735.

Duran, Leo. *Raw Silk: A Practical Handbook for the Buyer.* 2nd ed. New York, Silk Publishing Co., 1921.

Ebato Akira 江波戸昭. *Sanshigyō chiiki no keizai chirigakuteki no kenkyū* 蚕絲業地域の経濟地理學的の 研究(Research on the economic geography of sericultural districts). Tokyo, Kokon sho-in, 1969.

Elvin, Mark. *The Pattern of the Chinese Past.* Stanford, Stanford University Press, 1973.

Eng, Robert Y. "Silk Reeling Enterprises in Canton and Shanghai," paper presented at the Conference on the Colonial Port City in Asia, June 1976.

———. "Imperialism and the Chinese Economy: The Canton and Shanghai

Silk Industry, 1861–1932." PhD dissertation, University of California, Berkeley, 1978.

Fairbank, John K. *Trade and Diplomacy on the China Coast: The Opening of the Treaty Ports, 1842–1854.* Cambridge, Harvard University Press, 1953.

Fang-chih shih-hua 紡織史話 (Stories about the history of weaving), comp. Shang-hai shih fang-chih k'o-hsueh yen-chiu yuan 上海市紡織科學研究院 (Shanghai Weaving Science Research Institute). Shanghai, K'o-hsueh chi-shu ch'u-pan she, 1978.

Fei Hsiao-t'ung. *Peasant Life in China: A Field Study of Country Life in the Yangtze Valley.* London, Routledge and Kegan Paul, 1939.

Feuerwerker, Albert, ed. *History in Communist China.* Cambridge, Massachusetts Institute of Technology Press, 1968.

———. *The Chinese Economy, ca. 1870–1911.* Michigan Papers in Chinese Studies No. 5. Ann Arbor, University of Michigan, 1969.

———. "Handicraft and Manufactured Cotton Textiles in China, 1871–1910," *Journal of Economic History* 30.2:338–378 (June 1970).

Fong, H. D. "China's Silk Reeling Industry: A Survey of Its Development and Distribution," *Monthly Bulletin on Economic China* (Nankai Institute of Economics) 7.12:483–506 (December 1934).

Fortune, Robert. *Three Years' Wanderings in the Northern Provinces of China, Including a Visit to the Tea, Silk, and Cotton Countries.* London, 1847.

———. *Two Visits to the Tea Countries of China.* 2 vols. London, 1853.

———. *A Residence Among the Chinese: Inland, on the Coast, and at Sea.* London, 1857.

Foust, Clifford M. *Muscovites and Mandarins: Russia's Trade with China and its Setting, 1727–1805.* Chapel Hill, University of North Carolina Press, 1969.

Fu I-ling 傅衣凌. *Ming-Ch'ing shih-tai shang-jen chi shang-yeh tzu-pen* 明清時代商人及商業資本 (Merchants and commercial capital in the Ming and Ch'ing periods). Peking, Jen-min ch'u-pan she, 1956.

———. *Ming-tai Chiang-nan shih-min ching-chi shih-t'an* 明代江南市民経濟試探 (Theories about the economy of the urban populace of Kiangnan in the Ming dynasty). Shanghai, Jen-min ch'u-pan she, 1963.

Fu Lo-shu, comp. and trans. *A Documentary Chronicle of Sino-Western Relations.* 2 vols. Tucson, University of Arizona Press, 1966.

Fujimoto Jitsuya 藤本實也 . *Nihon sanshigyōshi* 日本蚕絲業史 (History of Japan's silk industry). 3 vols. n.p., 1933.

George, M. Dorothy. *England in Transition*. Rev. ed. Baltimore, Maryland, Penguin, 1953.

Han Ta-ch'eng 韓大成 . "Ming-tai shang-p'in ching-chi ti fa-chan yü tzu-pen chu-i ti meng-ya" 明代商品经済的發展與資本主義的萌芽 (The development of a commercial economy in Ming times and the "sprouts of capitalism") in *Ming-Ch'ing she-hui ching-chi hsing-tai ti yen-chiu* 明清社會经濟形態的研究 (Research on Ming and Ch'ing social and economic structure), comp. Chung-kuo jen-min ta-hsueh, Chung-kuo li-shih chiao-yen shih 中國人民大學,中國歷史教研室 (Chinese People's University, Department of Chinese History). Shanghai, Jen-min ch'u-pan she, 1956. Pp. 1–102.

Hang-chou fu chih 杭州府志 (Gazetteer of Hangchow prefecture). 1784.

Hang-chou shih ching-chi tiao-ch'a 杭州市经濟調查 (Investigation of the economy of Hangchow city), ed. Chien-she wei-yuan hui tiao-ch'a Che-chiang ching-chi so 建設委員會調查浙江经濟所 (Bureau for the Economic Survey of Chekiang, National Reconstruction Council). Hangchow, 1932.

Hao Yen-p'ing. *The Compradore in Nineteenth Century China*. Cambridge, Harvard University Press, 1970.

Hart, Robert. *The I.G. in Peking: Letters of Robert Hart, Chinese Maritime Customs, 1868–1907,* eds. John King Fairbank, Katherine Frost Bruner, and Elizabeth MacLeod Matheson. Cambridge, Harvard University Press, 1975.

Hatano Yoshihiro 波多野善大 . *Chūgoku kindai kōgyōshi no kenkyū* 中國近代工業史の研究 (Studies on the history of modern Chinese industry). Tokyo, Tōyōshi kenkyūkai, 1961.

Heard, Augustine, and Co. Archives. Baker Library, Harvard Business School.

Hemmi Kenzō. "Primary Product Exports and Economic Development: The Case of Silk," in Kazushi Ohkawa et al., eds., *Agriculture and Economic Growth: Japan's Experience*. Princeton, Princeton University Press, 1970. Pp. 303–323.

Hilton, Rodney, introduction. *The Transition from Feudalism to Capitalism*. London, New Left Books, 1976.

Hinder, Eleanor M. *Life and Labour in Shanghai: A Decade of Labour and Social Administration in the International Settlement.* New York, Institute of Pacific Relations, 1944.

Hirschmeier, Johannes. *The Origins of Entrepreneurship in Meiji Japan.* Cambridge, Harvard University Press, 1964.

Ho Ping-ti. *The Ladder of Success in Imperial China: Aspects of Social Mobility, 1368–1911.* New York, Science Editions, 1964.

—— 何炳棣 . *Chung-kuo hui-kuan shih-lun* 中國會館 史論 (A historical survey of *"Landsmannschaften"* in China). Taipei, T'ai-wan hsueh-sheng shu-chü, 1966.

Horie Eiichi 堀江英一 . *Shina sanshigyō ni okeru torihiki kankō: Keizai ni kansuru Shina kankō chōsa hōkokusho* 支那蚕絲業 における取引慣行 : 経済に関する支那 慣行調査報告書 (A report on the economic customs in China: On the practices in the Chinese silk industry). Tokyo, Tōa kenkyūjo, 1944.

Howard, Charles Walter and Karl P. Buswell. *A Survey of the Silk Industry of South China.* Hong Kong, Commercial Press, 1925.

Hsiao Liang-lin. *China's Foreign Trade Statistics, 1864–1949.* Cambridge, Harvard University, East Asian Research Center, 1974.

Hsu Kuang-ch'i 徐光啟 . *Nung-cheng ch'üan-shu* 農政全書 (Encyclopedia of agricultural techniques). 60 chüan. 1838 ed. Reprint, Peking, Chung-hua shu-chü, 1959.

Hu-chou fu chih 湖州府志 (Gazetteer of Hu-chou prefecture). 1874.

(Ch'in-ting) Hu-pu tse-li 欽定戸部則例 (Imperially Commissioned Regulations of the Board of Revenue). 100 chüan. 1865 ed.

Hu ts'an-shu 湖蚕述 (Story of sericulture in Hu-chou). Wang Yueh-chen 汪曰楨 . 1874. Reprint, Peking, Chung-hua shu-chü, 1956.

Huang, Ray. *Taxation and Governmental Finance in Sixteenth Century Ming China.* Cambridge, Cambridge University Press, 1974.

Huang-ch'ao ching-shih wen-pien 皇朝经世文編 (Collected essays on statecraft in the Ch'ing dynasty), comp. Ho Ch'ang-ling 賀長齡 . 120 chüan. 1886 ed.

Huber, Charles Joseph. *The Raw Silk Industry of Japan.* New York, Silk Association of America, 1929.

Hudson, G. F. *Europe and China.* London, Arnold, 1931.

Hummel, Arthur W., ed. *Eminent Chinese of the Ch'ing Period.* 2 vols.

Washington, D.C., Government Printing Office, 1943–1944. Reprint, Taipei, Ch'eng-wen, 1964.

Ishii Kanji 石井寛治 . *Nihon sanshigyōshi bunseki* 日本蚕絲業史分析 (Analysis of the history of Japanese sericulture). Tokyo, Tōkyō daigaku shuppansha, 1972.

Kawabata Genji 河鰭源治 . "Taihei tenkoku senryōka Nanjunchin ni okeru Koshi bōeki" 太平天國占領下南潯鎮 に お け る 湖絲貿易 (Hu-chou silk trade in Nan-hsun during the Taiping occupation), *Tōhōgaku* 東方學 (Oriental studies) 22:84–98 (July 1961).

Kitamura Hiranao 北村敬直 . "Shindai ni okeru Koshūfu Nanjinchin no mentonya ni tsuite" 清代における湖州府南潯鎮の棉間屋について (A yard goods store in Nan-hsun *chen*, Hu-chou fu, during the Ch'ing period), *Keizaigaku zasshi* 経濟學雜誌 (Journal of economic studies) 57.3:1–19 (September 1967).

Ko-sheng-ch'ü li-nien ts'ai-cheng hui-lan 各省區歷年財政彙覽 (Annual compendium of the financial administration of provinces and regions), comp. Ts'ai-cheng pu, ts'ai-cheng tiao-ch'a ch'u 財政部財政調查處 (Ministry of Finance, Financial Administration Investigation Bureau). 4 ts'e. Peking, 1927.

Kojima Yoshio 小島淑男 . "Shimmatsu minkoku-shōki Soshūfu no kinuorigyō to kiko no dōkō" 清末民國初期蘇州府の絹織業と機户の動向 (The silk industry and weavers' movement in Soochow prefecture in the late Ch'ing and early Republican period), *Shakai keizai shigaku* 社會経濟史學 (Socioeconomic history) 34.5:32–54 (January 1969).

Kōsoshō Mushakuken nōson jittai chōsa hōkokusho 江蘇省無錫縣農村實態調查報告書 (Report on the investigation of rural villages in Wusih hsien, Kiangsu province), comp. Minami Manshū tetsudō kabushiki kaisha, Shanhai jimusho chōsashitsu 南滿洲鐵道株式會社上海事務所調查室 (The Shanghai Investigation Bureau of the South Manchurian Railway Co.). Shanghai, 1941.

Kuang ts'an-sang shuo chi-pu 廣蚕桑説輯補 (Expanded edition of the *Ts'an-sang shuo*), comp. Chung Hsueh-lu 仲學輅 . Rev. ed. 1877.

Kuei-an hsien chih 歸安縣志 (Gazetteer of Kuei-an hsien). 1882.

"Kwangtung Silk Worms," *Chinese Economic Journal* 5.2:714–728 (August 1929).

Li Chih-chin 李之勤 . "Lun ya-p'ien chan-cheng i-ch'ien Ch'ing-tai shang-yeh-hsing nung-yeh ti fa-chan" 論雅片戰爭以前清代商業性農業的發展 (On the development of commercialized agriculture in the Ch'ing dynasty before the Opium War) in *Ming-Ch'ing she-hui ching-chi hsing-t'ai ti yen-chiu* 明清社會經濟形態的研究 (Research on Ming and Ch'ing social and economic structure), comp. Chung-kuo jen-min ta-hsueh, Chung-kuo li-shih chiao-yen shih (Chinese People's University, Department of Chinese History). Shanghai, Jen-min ch'u-pan she, 1956. Pp. 263–357.

Li Hsu 李煦 . *Su-chou chih-tsao Li Hsu tsou-che* 蘇州織造李煦奏摺 (Memorials of the Soochow textile commissioner Li Hsu). 1 ts'e. Peking, 1937.

Li, Lillian Ming-tse. "Kiangnan and the Silk Export Trade, 1842–1937." PhD dissertation, Harvard University, 1975.

——. "The Silk Export Trade and the Modernization of China and Japan," a paper presented at the Conference on American-East Asian Relations, June 1976.

Li Shan-shu 李善述 . *K'ao-ch'a Ou-Mei sheng-ssu shih-ch'ang pao-kao* 考察歐美生絲市場報告 (Report on an investigation of the European and American raw silk markets). Soochow, June 1937.

Lieu, D. K. *The Silk Industry of China.* Shanghai, Kelly and Walsh, 1941.

Liu Yung-ch'eng 劉永成 . "Shih-lun Ch'ing-tai Su-chou shou-kung-yeh hang-hui" 試論清代蘇州手工業行會 (Essay on the handicraft guilds of Soochow during the Ch'ing dynasty), *Li-shih yen-chiu* 歷史研究 (Historical research) 1959.11:21–46 (November 1959).

Lockwood, William W. *The Economic Development of Japan.* Rev. ed. Princeton, Princeton University Press, 1968.

Lopez, R. S. "China Silk in Europe in the Yuan Period," *Journal of the American Oriental Society* 72.2:72–76 (April-June 1952).

Mao Tun. *Spring Silkworms and Other Stories.* Peking, Foreign Languages Press, 1956.

Markham, Jesse W. *Competition in the Rayon Industry.* Cambridge, Harvard University Press, 1952.

McElderry, Andrea Lee. *Shanghai Old-Style Banks (Ch'ien-chuang), 1800–1935.* Michigan Papers in Chinese Studies No. 25. Ann Arbor, University of Michigan, 1976.

Milburn, William. *Oriental Commerce.* 2 vols. London, Black, Parry & Co., 1813.

Morita Akira　森田明. *Shindai suirishi kenkyū* 清代水利史研究 (Studies on the history of water conservancy in the Ch'ing dynasty). Tokyo, Akishobō, 1974.

Morse, Hosea Ballou. *The International Relations of the Chinese Empire.* 3 vols. London, Longmans, Green, 1910–1918.

———. *The Chronicles of the East India Company Trading to China, 1635–1834.* 5 vols. Oxford, Clarendon Press, 1926.

Moulder, Frances V. *Japan, China and the modern world economy.* Cambridge, Cambridge University Press, 1977.

Murphey, Rhoads. *Shanghai: Key to Modern China.* Cambridge, Harvard University Press, 1953.

———. *The Treaty Ports and China's Modernization: What Went Wrong?* Michigan Papers in Chinese Studies No. 7. Ann Arbor, University of Michigan, 1970.

Mushaku no seishigyō 無錫の製絲業 (The silk industry of Wusih), ed. Minami Manshū tetsudō kabushiki kaisha, Shanhai jimusho chōsashitsu 南滿洲鐵道株式會社上海事務所調査室 (The Shanghai Investigation Bureau of the South Manchurian Railway Co.). Shanghai, 1940.

Nakamura Masanori 中村正則. "Seishigyō no tenkai to jinushisei" 製絲業の展開と地主制 (The landlord system and the development of the silk reeling industry), *Shakai keizai shigaku* 社會経濟史學 (Studies in social and economic history) 32.5 & 6: 46–71 (1967).

Nan-hsun chih 南潯志 (Gazetteer of Nan-hsun). 1922.

Negishi Tadashi 根岸佶. *Shanhai no girudo* 上海のギルド (The guilds of Shanghai). Tokyo, Nihon hyōronsha, 1951.

Nihon boeki seiran 日本貿易精覽 (The foreign trade of Japan: A statistical survey), comp. Tōyō keizai shimpōsha 東洋経濟新報社 (Oriental economic news survey). Tokyo, 1935, reprinted 1975.

Nihon sangyōshi taikei 日本産業史体系 (Outline of the history of Japanese industry), comp. Chihōshi kenkyū kyōgikai 地方史研究協議會 (Research Conference on Local History). 9 vols. Tokyo, Tōkyō daigaku shuppansha, 1960–1961.

Nihon sen'i sangyōshi 日本纖維産業史 (History of Japanese textile production), comp. Nihon sen'i kyōgikai 日本纖維協議會 (Japan Textile Council). 2 vols. Osaka, 1958.

Nung-hsueh pao 農學報 (Journal of agricultural studies). Shanghai, 1895–1905.

Nung-sang chi-yao 農桑輯要 (The essentials of agriculture and sericulture), comp. Yuan ssu-nung ssu 元司農司 (Agriculture Ministry, Yuan dynasty). 7 chüan. In *Wu-ying tien chü chen pan ch'üan-shu* 武英殿聚珍版全書. 800 ts'e. 1899 ed.

Nung-shang t'ung-chi piao 農商統計表 (Statistical tables on agriculture and commerce), comp. Chung-kuo nung-shang pu, tsung-wu t'ing, t'ung-chi k'o 中國農商部總務廳統計科 (Statistical division of the general office of the Ministry of Agriculture and Commerce). Peking, 1914–1924.

Nung-shu 農書 (Book of agriculture). Ch'en Fu 陳敷. 1154 ed. In *Chih pu-tsu chai ts'ung-shu* 知不足齋叢書. 240 ts'e. 1921 ed.

Nung-shu 農書 (Book of agriculture). Wang Chen 王楨. 22 chüan. 1924 ed.

Ohara Keishi, comp. *Japanese Trade and Industry in the Meiji-Taisho Era.* Tokyo, Ōbunsha, 1957.

Pasteur, Louis. *Etudes sur la maladie des vers à soie.* 2 vols. Paris, Gauthier-Villars, 1870.

Patrick, Hugh T. "Japan, 1868–1914," in Rondo E. Cameron, ed., *Banking in the Early Stages of Industrialization.* New York, Oxford University Press, 1967.

P'eng Tse-i 彭澤益. "Ts'ung Ming-tai kuan-ying chih-tsao ti ching-ying fang-shih k'an Chiang-nan ssu-chih-yeh sheng-ch'an ti hsing-chih" 從明代官營織造的经营方式看江南絲織業生產的性質 (Looking at the nature of the Kiangnan silk weaving industry's production from the commercial practices of the Ming dynasty government-sponsored silk weaving factories), *Li-shih yen-chiu* 1963.2:33–56 (February 1963).

——. "Ch'ing-tai ch'ien-ch'i Chiang-nan chih-tsao ti yen-chiu" 清代前期江南織造的研究 (Research on the Kiangnan silk weaving factories in the early Ch'ing period), *Li-shih yen-chiu* 1963.4:91–116 (April 1963).

——. "Ya-pien chan-cheng ch'ien Ch'ing-tai Su-chou ssu-chih-yeh sheng-ch'an kuan-hsi ti hsing-shih yü hsing-chih" 雅片戰爭前清代蘇州絲織業生產關係的形式與性質 (The structure and nature of the productive relations in the Soochow silk weaving industry in the Ch'ing dynasty before the Opium War), *Ching-chi yen-chiu* 经濟研究 (Economic research) 10:63–73 (October 1963).

——. "Shih-chiu shih-chi hou-ch'i, Chung-kuo ch'eng-shih shou-kung-yeh shang-yeh hang-hui ti ch'ung-chien ho tso-yung" 十九世紀後期中國城市手工業商業行會的重建和作用 (The reconstruction and functions of the Chinese urban handicraft and commercial guilds in the late nineteenth century), *Li-shih yen-chiu* 1965.1:71–102 (January 1965).

Perkins, Dwight H. *Agricultural Development in China, 1368–1968.* Chicago, Aldine, 1969.

P'ing-hu hsien chih 平湖縣志 (Gazetteer of P'ing-hu hsien). 1886.

Pritchard, Earl H. *The Crucial Years of the Early Anglo-Chinese Relations, 1750–1800.* Pullman, Washington, State College of Washington Press, 1936.

Pu Nung-shu 補農書 (The *Nung-shu* amended), ed. and amended by Chang Lü-hsiang 張履祥. 1897. Reprinted as *Shen-shih Nung-shu* 沈氏農書 (Mr. Shen's *Nung-shu*). Peking, Chung-hua shu-chü, 1956.

Ramsay, George David. *The Wiltshire Woollen Industry in the Sixteenth and Seventeenth Century.* 2nd ed. London, F. Cass, 1965.

Remer, C. F. *The Foreign Trade of China.* Shanghai, Commercial Press, 1926.

Rice, Eugene F., Jr. *The Foundations of Early Modern Europe, 1460–1559.* New York, W. W. Norton, 1970.

Riskin, Carl. "Surplus and Stagnation in Modern China," in Dwight H. Perkins, ed., *China's Modern Economy in Historical Perspective.* Stanford, Stanford University Press, 1975. Pp. 49–82.

Rondot, Natalis. *Les soies.* 2 vols. 2nd ed. Paris, 1885–1887.

Saeki Yūichi 佐伯有一 . "Min zenpanki no kiko" 明前半期 の機戸 (Silk weavers during the first half of the Ming), *Tōyō bunka kenkyūjo kiyō* 東洋文化研究所紀要 (Memoirs of the Institute for Oriental Culture) 8:167–210 (March 1956).

———. "Mindai shōekisei no hōkai to toshi kinuorimonogyō ryūtsūshijō no tenkai" 明代匠役制の崩壊と都市絹織物 業の流通市場の展開 (The collapse of the system of government-employed artisans and the development of silk markets in Chinese cities), *Tōyō bunka kenkyūjo kiyō* 10:359–425 (November 1956).

———. "Sen-roppyaku-ichi-nen 'shokuyō no hen' o meguru shomondai" 一六○一年'織傭之變'をめぐる諸問題 (Problems of the silk weavers' revolt of 1601 in Soochow), *Tōyō bunka kenkyūjo kiyō* 45:77–108 (March 1968).

Sargent, A. J. *Anglo-Chinese Commerce and Diplomacy.* Oxford, Oxford University Press, 1907.

Schurz, William Lytle. *The Manila Galleon.* New York, E. P. Dutton, 1939.

"Sericulture in Wusih," *Chinese Economic Journal* 1.2:142–159 (February 1927).

Serruys, Henry. "Sino-Mongol Relations during the Ming, II: The Tribute System and Diplomatic Missions (1400–1600)," *Mélanges chinois et bouddhiques* 14:1–650 (1967).

Shanghai International Testing House. *China Silk.* Shanghai, 1925.

"Shanghai Silk Filatures," *Chinese Economic Journal* 3.1:590–604 (July 1928).

Shiba Sakuo 柴三九男 . "Shimmatsu Kanton sankakusu no yōsan keiei to nōson kindaika – Tōyōteki shakai to 'gyotō'" 清末廣東 三角州の養蚕経営と農村近代化—東洋的 社會と'魚塘' (Sericultural enterprise in the Canton delta in the late Ch'ing and village modernization—Oriental society and "yü-t'ang"), *Shikan* 史觀 (The historical review) 57–58:22–34 (March 1960).

Shiba Yoshinobu. *Commerce and Society in Sung China,* trans. Mark Elvin. Michigan Abstracts of Chinese and Japanese Works on Chinese History. Ann Arbor, University of Michigan, 1970.

Shih-men hsien chih 石門縣志 (Gazetteer of Shih-men hsien). 1879.

Shih Min-hsiung 施敏雄 . *Ch'ing-tai ssu-chih kung-yeh ti fa-chan* 清 代絲織工業的發展 (The development of the silk manufacturing industry of the Ch'ing dynasty). Taipei, Commercial Press, 1968.

Shina keizai zensho 支那経濟全書 (Encyclopedia of the Chinese economy), comp. Tōa dōbunkai 東亞同文會 (East Asian Cultural Institute). 12 vols. Osaka and Tokyo, 1907–1908.

Shina sanshigyō kenkyū 支那蚕絲業研究 (Research on the Chinese silk industry), comp. Tōa kenkyūjo 東亞研究所 (Oriental Research Institute). Osaka, Osakagyō shōten, 1943.

Shina sanshigyō taikan 支那蚕絲業大觀 (A general survey of the silk industry in China), comp. Sanshigyō dōgyōkumiai chūōkai 蚕絲業同業组合中央會 (Central Association of the Cooperative Union of the Silk Industry). Tokyo, Okada nichieidō, 1929.

Shina seisan jigyō tōkeihyō 支那生產實業統計表 (Statistical tables of China's production), comp. Shinkoku nōkōshōbu 清國農工商部 (China, Ministry of Agriculture, Industry, and Commerce). Japanese reprint. 2 vols. Osaka, 1912.

Shina shōbetsu zenshi 支那省別全志 (Encyclopedia of the Chinese provinces), comp. Tōa dōbunkai. 18 vols. Tokyo, 1917–1920.

Shinohara Miyohei. "Economic Development and Foreign Trade in Pre-War Japan," in C. D. Cowan, ed., *The Economic Development of China and Japan.* New York and London, Frederick A. Praeger, 1964.

Shou-shih t'ung-k'ao 授時通考 (Encyclopedia of agriculture), ed. Chiang P'u 蔣溥 . 1742. Reprint, Peking, Chung-hua shu-chü, 1956.

Shuang-lin chen chih 雙林鎮志 (Gazetteer of Shuang-lin *chen*). 1880.

Shun-te hsien chih 順德縣志 (Gazetteer of Shun-te hsien). 1929.

Silk, comp. China, Imperial Maritime Customs. Special Series. Shanghai, 1881.

"The Silk Export Trade of China," *Chinese Economic Journal* 3.5:942–964 (November 1928).

"The Silk Industry in Kwangtung Province," *Chinese Economic Journal* 5.1:604–620 (July 1929).

Silk: Replies from the Commissioners of Customs to the Inspector General's Circular no. 103, to which is added "Manchuria Tussore Silk," comp. China, Maritime Customs. Shanghai, 1917.

Smelser, Neil J. *Social Change in the Industrial Revolution: An Application of Theory to the British Cotton Industry.* Chicago, University of Chicago Press, 1959.

Smith, Thomas C. *Political Change and Industrial Development in Japan: Government Enterprise, 1868–1880.* Stanford, Stanford University Press, 1955.

Spence, Jonathan D. *Ts'ao Yin and the K'ang-hsi Emperor: Bondservant and Master.* New Haven, Yale University Press, 1966.

Ssu-yeh chih yu 絲業之友 (Friends of the silk industry). Special issue in memory of Li Yu-jen 李佑仁. 28 (December 5, 1940).

Stanley, C. John. *Late Ch'ing Finance: Hu Kuang-yung as an Innovator.* Cambridge, Harvard University, East Asian Research Center, 1961.

Sun Ching-chih, et al., eds. *Economic Geography of the East China Region.* Translation of *Hua-tung ti-ch'ü ching-chi ti-li* 華東地區經濟地理. Washington, D.C., Joint Publications Research Service, 1961.

Sun, E-tu Zen. "Sericulture and Silk Textile Production in Ch'ing China," in W. E. Willmott, ed., *Economic Organization in Chinese Society.* Stanford, Stanford University Press, 1972. Pp. 79–108.

Sung Ying-hsing 宋應星. *T'ien-kung k'ai-wu: Chinese Technology in the Seventeenth Century,* trans. E-tu Zen Sun and Shiou-chuan Sun. University Park and London, Pennsylvania State University Press, 1966.

A Survey of the Silk Industry of Central China, comp. Shanghai International Testing House. Shanghai, 1925.

Suzuki Tomoo 鈴木智夫. "Shimmatsu Minsho ni okeru minzoku shihon no tenkai katei: Kanton no seishigyō ni tsuite," 清末民初に
おける民族資本の展開過程：廣東の生絲
業について (The process of development of "native capital" during the late Ch'ing and the early Republic), in *Chūgoku kindaika no shakai kōzō* 中國近代化の社會構造 (Social structure of the modernization of China), in *Tōyō shigaku ronshū* 東洋史學論集 (Collection on East Asian historiography), Vol. 6. Tōkyō kyōiku daigaku bungakubu. 1960, reprinted by Daian, 1966. Pp. 45–70.

(Ch'in-ting) Ta-Ch'ing hui-tien shih-li 欽定大清會典事例 (Precedents pertaining to the collected statutes of the Ch'ing dynasty). 1220 chüan. Kuang-hsu ed.

Taikei Nihonshi sōsho 体系日本史叢書 (Outline of Japanese history), comp. Yamakawa shuppansha. 24 vols. Tokyo, 1964–1970.

(Hsu) Tan-t'u hsien chih 續丹徒縣志 (Gazetteer of Tan-t'u hsien continued). 1930.

Tanaka Masatoshi 田中正後 . "'Shihon shugi no hōga' kenkyū" '資本主義 の 萌芽'研究 (Researches on the "sprouts of capitalism"), in his *Chūgoku kindai keizaishi kenkyū josetsu* 中國 近代経濟史研究序説 (An introduction to the study of the economic history of modern China). Tokyo, Tokyo University Press, 1973. Pp. 205–241.

Te-ch'ing hsien hsin-chih 德清縣新志 (New gazetteer of Te-ch'ing hsien). 1931.

Terada Takanobu 寺田隆信 . "Soshū chihō ni okeru toshi no mengyō shōnin ni tsuite" 蘇松地方 に 於 ける 都市 の 棉業商人 に つ い て (Cotton merchants in urban markets in Soochow and Sungkiang) *Shirin* 史林 (Journal of history) 41.6:52–69 (1958).

Ting Yueh-hung C. "Sericulture in Hu-chou as seen in the *Hu-chou fu-chih*," *Papers on China,* Harvard University, East Asian Research Center, 23:29–51 (July 1970).

Topley, Margaret. "Marriage Resistance in Rural Kwangtung," in Margery Wolf and Roxane Witke, eds., *Women in Chinese Society*. Stanford, Stanford University Press, 1975. Pp. 67–88.

Ts'an-sang chien-ming chi-shuo 蠶桑簡明輯説 (Simplified explanation of sericulture). Huang Shih-pen 黄世本 . 1 ts'e. Rev. ed. 1888.

Ts'an-sang ts'ui pien 蠶桑萃編 (Collected materials on sericulture). Wei Chieh 衛杰 . 13 chüan. 1900. Peking reprint, 1956.

Ts'an-shih yao-lueh 蠶事要略 (Essentials of sericulture). By Chang Hsing-fu 張行孚 . Original date unclear. In Yuan Ch'ang 袁昶 , *Chien-hsi ts'un-she ts'ung-k'o* 漸西村舎叢刻 (Collected works from Chien-hsi). 66 ts'e. 1895 ed.

Ts'an-shu 蠶書 (Book of sericulture). Ch'in Kuan 秦觀 . In *Chih pu-tsu chai ts'ung-shu* 知不足齋叢書 . 240 ts'e. 1921 ed.

Ts'an-wu shuo-lueh 蠶務説略 (Brief explanation of sericulture), ed. K'ang Fa-ta 康發達 (F. Kleinwachter). n.d. In *Chih-hsueh ts'ung-shu ch'u-chi* 質學叢書初集 . 38 ts'e. 1897 ed.

Tseng T'ung-ch'un 曾同春 . *Chung-kuo ssu-yeh* 中國絲業 (The Chinese silk industry). Shanghai, Commercial Press, 1933.

T'ung-hsiang hsien chih 桐鄉縣志 (Gazetteer of T'ung-hsiang hsien). 1881.

Uchida Hoshimi 内田星美 . *Nihon bōshoku gijitsu no rekishi* 日本

紡織技術の歷史 (History of Japanese textile technology). Tokyo, Chijinshukan, 1960.

Union des Marchands de Soie de Lyon. *Statistique de la production de la soie en France et l'étranger.* Lyon, 1880–1923 annual.

Wang Chung-lo 王仲犖 . "Ming-tai Su-Sung-Chia-Hu ssu-fu ti tsu-o ho Chiang-nan fang-chih yeh" 明代蘇松嘉湖四府的租額和江南紡織業 (The tax quotas of the four prefectures of Su-Sung-Chia-Hu during the Ming and the Kiangnan textile industry), in *Chung-kuo tzu-pen chu-i meng-ya wen-t'i t'ao-lun chi* 中國資本主義萌芽問題討論集 (Essays on the question of the "sprouts of capitalism" in China), ed. Chung-kuo jen-min ta-hseuh, Chung-kuo li-shih chiao-yen shih (Chinese People's University, Department of Chinese History). Peking, San-lien shu-tien, 1957. Pp. 1–19.

Wang Yeh-chien. "The Impact of the Taiping Rebellion on the Population in Southern Kiangsu," *Papers on China,* Harvard University, East Asian Research Center, 19:120–158 (December 1965).

———. "The Secular Trend of Prices during the Ch'ing Period (1644–1911)," *Journal of the Institute of Chinese Studies of the Chinese University of Hong Kong* 5.2:347–371 (1972).

Wang Yü-hu 王毓瑚. *Chung-kuo nung-hsueh shu-lu* 中國農學書錄 (Bibliography of Chinese agricultural works). Peking, Chung-hua shu-chü, 1957.

Wiens, Mi Chu. "Socioeconomic Change during the Ming Dynasty in the Kiangnan Area." PhD dissertation, Harvard University, 1973.

Williams, S. Wells. *The Chinese Commercial Guide.* 5th ed. Hong Kong, A. Shortrede, 1863.

———. *The Middle Kingdom.* Rev. ed. 2 vols. New York, Charles Scribner, 1883. Reprint, Taipei, Ch'eng-wen, 1965.

Wills, John E., Jr. "Ch'ing Relations with the Dutch, 1662–1690," in John King Fairbank, ed., *The Chinese World Order: Traditional China's Foreign Relations.* Cambridge, Harvard University Press, 1968. Pp. 225–256.

Wu-ch'eng hsien chih 烏程縣志 (Gazetteer of Wu-ch'eng hsien). 1881.

Wu Ch'eng-lo 吳承洛 . *Chung-kuo tu-liang heng shih* 中國度量衡史 (A history of Chinese weights and measures). Shanghai, Commercial Press, 1937.

Wu-ch'ing chen chih 烏青鎮志 (Gazetteer of Wu-ch'ing *chen*). 1760 and 1936.

Wu-hsi Chin-kuei hsien chih 無錫金匱縣志 (Gazetteer of Wu-sih and Chin-kuei hsiens). 1881.

Wu-hsing nung-ts'un ching-chi 吳興農村経濟 (The rural economy of Wu-hsing), comp. Chung-kuo ching-chi t'ung-chi yen-chiu so 中國経濟统計研究所 (Research Bureau on Chinese Economic Statistics). Shanghai, 1939.

Wu-hsu pien-fa 戊戌變法 (The 1898 reform movement), ed. Chung-kuo shih-hsueh hui 中國史學會 (Chinese Historical Society). Shanghai, Shen-chou kuo-kuang she, 1953.

Wu-hsu pien-fa tang-an shih-liao 戊戌變法檔案史料 (Archives of the 1898 reform movement), comp. Kuo-chia tang-an chü, Ming-Ch'ing tang-an kuan 國家檔案局,明清檔案館 (National Archival Bureau, Ming-Ch'ing Archival Office). Peking, Chung-hua shu-chü, 1958.

Yagi Haruo 矢木明夫 "Seishigyō" 製絲業 (The silk manufacturing industry), in *Nihon sangyōshi taikei* 日本產業史大系 (Outline of the history of Japanese manufacturing), comp. Chihōshi kenkyū kyōgikai. Tokyo, Tōkyō daigaku shuppansha, 1961. I, 221–244.

———. *Nihon keizaishi gaisetsu* 日本経濟史概説 (Outline of Japanese economic history). Tokyo, Hyōronsha, 1974.

Yamaguchi Kazuo 山口和雄 ed. *Nihon sangyōkin'yūshi kenkyū: seishi kin'yūhen* 日本產業金融史研究 – 製絲金融篇 (Research on the history of Japanese industrial finance: section on silk manufacturing finance). Tokyo, Tōkyō daigaku shuppansha, 1966.

Yanagida Setsuko 柳田節子 . "Sōdai no yōsan nōka keiei: Kōnan o chūshin to shite" 宋代の養蚕農家経營：江南を中心として (Peasant sericultural enterprise in the Sung dynasty: Kiangnan), in *Wada Hakushi koki kinen Tōyōshi ronsō* 和田博士古稀纪念東洋史論叢 (*Oriental Studies* presented to Wada Sei in celebration of his seventieth birthday) comp. Wada Hakushi koki kinen Tōyōshi ronsō hensan iinkai 編纂委員會 (Committee for editing *Oriental Studies* commemorating the seventieth birthday of Wada Sei). Tokyo, Kodansha, 1961. Pp. 993–1003.

Yang Lien-sheng. "The Concept of 'Pao' as the Basis for Social Relations in China," in John K. Fairbank, ed., *Chinese Thought and Institutions*. Chicago, University of Chicago Press, 1957. Pp. 291–301.

———. "Government Control of Urban Merchants in Traditional China," *Tsinghua Journal of Chinese Studies* (Ch'ing-hua hsueh-pao 清華 學報), new series, 8.1 and 2:186–209 (August 1970). In English with a Chinese summary.

Yang Ta-chin 楊大金 . *Hsien-tai Chung-kuo shih-yeh chih* 現代 中國實業志 (Gazetteer of modern Chinese industry). Rev. ed. 2 vols. Changsha, 1938.

Yang Tuan-liu (C. Yang) and Hou-pei Hou (H. B. Hau), comp. *Statistics of China's Foreign Trade during the Last Sixty-five Years, 1864–1928.* Nanking, Academia Sinica, 1931.

Yen Chung-p'ing 嚴中平 . "Ming-Ch'ing liang-tai ti-fang kuan ch'ang-tao fang-chih yeh shih-li" 明清兩代地方官倡導紡織 業示例 (Examples of the promotion of spinning and weaving industry by local officials in the Ming and the Ch'ing), *Tung-fang tsa-chih* 東方雜誌 (Eastern miscellany) 42.8:20–26 (April 1946).

———, et al., comp. *Chung-kuo chin-tai ching-chi shih t'ung-chi tzu-liao hsuan-chi* 中國近代經濟史統計資料選輯 (A selection of statistical materials on modern Chinese economic history). Peking, K'o-hsueh ch'u-pan she, 1955.

Yin Liang-ying 尹良瑩 . *Chung-kuo ts'an-yeh shih* 中國蠶業 史 (History of the Chinese silk industry). Nanking, National Central University, 1931.

Yokoyama Suguru 橫山英 . "Shindai no toshi kinuorimonogyō no seisan keitai" 清代の都市絹織物業の生産 形態. (The form of production in the urban silk weaving industry of the Ch'ing dynasty), *Shigaku kenkyū* 史學研究 (Historical research) 104:67–78, 105:52–66 (1968).

———. "Tampugyō no seisan kōzō," 端布業の生産構造 (The productive structure of the cotton sizing industry), in his *Chūgoku kindaika no keizai kōzō* 中國近代化の経濟構造 (The economic structure of Chinese modernization). Tokyo, Akishobo, 1972. Pp. 63–146.

Yokoyama Tadao. *Synthesized Science of Sericulture.* Bombay, Central Silk Board, 1962.

Yoshida Kazuko 吉田和子 . "Meiji shoki no seishi gijitsu ni okeru dochaku to gairai—Joshu no baai to Shinshu no baai" 明治初期の 製絲技術における土着と外来—上州の場合と 信州の場合 (The problem of native and foreign technologies in

the Joshu and Shinshu silk industries of the early Meiji period),
Kagakushi kenkyū 科學史研究　　　(Research on the history of
science) 2.16:16–24 (Spring 1977).

Yü-chih keng-chih t'u 御製耕織圖 (Imperially commissioned
illustrations of farming and weaving), comp. Chiao Ping-chen 焦秉
貞 . 1696 ed.

Yü Ying-shih. *Trade and Expansion in Han China: A Study of the Structure of Sino-Barbarian Economic Relations.* Berkeley and Los Angeles,
University of California Press, 1967.

Yueh Ssu-ping 樂嗣炳. *Chung-kuo ts'an-ssu* 中國蠶絲 (Chinese sericulture). Shanghai, Shih-chieh shu-chü, 1935.

Glossary I

Chinese and Japanese Names and Terms

Amoy (Hsia-men) 廈門
An-chi 安吉
An-tung 安東
Anhwei (An-hui) 安徽

Canton (Kuang-chou) 廣州
Chang Chien 張謇
Chang Chih-tung 張之洞
Chang-chou 漳州
chang-fang 賬房
Ch'ang-an 常安
Ch'ang-chou 常州
　(prefecture)
Ch'ang-chou 長州
　(hsien)
Ch'ang-hsing 長興
Ch'ang-hua 昌化
Ch'ang-shu 常熟
che 柘
ch'e-chiang 車匠
Chefoo (Yen-t'ai) 烟台
Chekiang (Che-chiang) 浙江
chen 鎮
Chen-ch'ang 震昌
Chen-chiang 鎮江
Chen I 振藝
Chen-t'ai 振泰
Chen-tse 震澤

Ch'en Ch'i-yuan 陳啟沅
Ch'eng-hua 成化
ch'eng-kuan 承管
Ch'eng-tu 成都
chi-chiang 機匠
chi-hu 機戶
Chi-yeh kung-so 機業公所
Ch'i-ch'ang ssu-ch'ang
　旗昌絲廠
Ch'i-hsiang kung-so 七襄公所

Ch'i-li 七里
Chia-hsing 嘉興
chia-sang 家桑
Chia-shan 嘉善
Chia-ting 嘉定
Chiang-ning 江寧
Chiang-p'u 江浦
chien-chan 繭棧
Chien-sheng 乾甡
chien-shih 繭市
ch'ien 錢
ch'ien-chuang 錢莊
Ch'ien-lung 乾隆
Ch'ien-t'ang 錢塘
chih-tsao chü 織造局
ch'ih 尺
Chihli 直隸

chin (catty) 斤
Chin-chi 錦記
Chin-kuei 金匱
chin-shih 進士
Ch'in 秦
ching 經
Ching-chiang 靖江
ching-hang 經行
Ching-pang ching-yeh kung-so
　　京幫經業公所
Ching-sang 荊桑
ching-shih 經世
Ching-yeh kung-so 經業公所

ching-ssu-hang 經絲行
ch'ing-ming 清明
ch'ing-sang yeh-hang
　　青桑葉行
Ch'ing-yuan 清遠
Chinkiang (Chen-chiang) 鎮江
Ch'iu Min-t'ing 邱敏庭
chou 州
Chou Hsiang-ling 周湘舲
Chu-chi 諸暨
Chu Pao-san 朱葆三
chu-tso chiang 住坐匠
chü-jen 舉人
Chü-jung 句容
chuang 莊
Chung-shan 中山
Chung-shan ko 中山葛
Ch'ung-te 崇德

erh-shu 二叔

Fei Hsiao-t'ung 費孝通
fei-ssu 肥絲
fen 分
Fo-shan 佛山
Foochow (Fu-chou) 福州
fu 府
Fu Hsi 伏羲

Fu-hwa 福華
Fu-yang 富陽
Fukien 福建
Fukushima 福島

Gifu 岐阜
Gumma 群馬
Gunze 郡是

Hai-men 海門
Hai-ning 海寧
Hai-yen 海鹽
Han 漢
Hangchow (Hang-chou)
　　杭州
Hankow (Han-k'ou)
　　漢口
hao 號
Heng-li 恆利
Ho-shan 鶴山
Hou-sheng 厚生
Hsia-chang kung-so
　　霞章公所
hsiang 鄉
hsiang-ching 鄉經
Hsiang-shan 香山
hsiang-ssu hang 鄉絲行
hsiang-tung 鄉董
Hsiao-feng 孝豐
hsiao ling-t'ou 小零頭
hsiao-pao 小包
Hsiao-shan 蕭山
hsien 縣
Hsien-feng 咸豐
hsien-shao 現稍
Hsin-ch'ang 新昌
Hsin-ch'eng 新城
Hsin-hui 新會
Hsiu-shui 秀水
Hsiung-nu 匈奴
Hsu Ping-k'un 許炳坤
Hsuan-[te] 宣德

Hsueh Fu-ch'eng 薛福成
Hsueh Hao-feng 薛浩峰
Hsueh Nan-mo 薛南溟
Hsueh Shou-hsuan 薛壽萱
Hu Ch'eng-mou 胡承謀
Hu-chou 湖州
Hu Kuang-yung 胡光鏞
Hu pu 戶部
Hu-sang 湖桑
Hu-shu kuan nü-tzu hsueh-hsiao
　　滸墅關女子學校
Hu-ts'ou 湖綢
hua-lou chi 花樓機
huan-tai 喚代
Huang Chi-wen 黃吉文
Huang-hsi 黃溪
Huang-ts'e 黃冊
Huang Tso-ch'ing 黃佐卿
hui-kuan 會館
Hunan 湖南
[Hung-] hsi 洪熙
Hupei 湖北

I-ho ssu-ch'ang 怡和絲廠
I-hsing 宜興

Jen-ho 仁和
Joshu 上卅
Jui-Lun 瑞倫
Jung-yeh kung-so 絨業公所

K'ai-hsien kung 開弦弓
kairyō zaguri ito
　　改良座繰絲
K'ang-hsi 康熙
Kansai 關西
Katakura 片倉
Kiangnan (Chiang-nan) 江南
Kiangsi (Chiang-hsi) 江西
Kiangsu (Chiang-su) 江蘇
ko 箇
Ko Ch'eng 葛城

k'o-shang 客商
Ku Yen-wu 顧炎武
K'uai-chi 會稽
Kuan-ch'eng t'ang 觀成堂
kuan-chiang 官匠
kuan-chüan 官絹
Kuang-ching 廣經
Kuang-hang 廣行
Kuang-hsu 光緒
Kuei-an 歸安
kung-chiang 工匠
Kung-ho yung 公和永
kung-so 公所
kung-ssu 貢絲
Kwangsi 廣西
Kwangtung 廣東

lan-chih 攬織
li 里
li-chin (likin) 釐金
li-hsia 立夏
Li Hung-chang 李鴻章
Li-yang 溧陽
liang 兩
Liang Ch'i-ch'ao 梁啟超
Liang-Kiang (Liang-chiang)
　　兩江
liao-ching 料經
Lin-an 臨安
Lin-hsi fang 濂溪坊
Lin Ti-ch'en 林迪臣
ling-chi kei-t'ieh
　　領機給帖
ling-chih 領織
Ling-hu 菱湖
ling-t'ou 零頭
Liu K'un-i 劉坤一
Lo Chen-yü 羅振玉
Lou Shou 樓璹
Lu-sang 魯桑
lun-pan chiang 輪班匠

273

mai-ssu chao-pan 買絲招辦
Mao Tun 茅盾
Meiji 明治
miao-sang 秒桑
miao-yeh 秒葉
mien-ch'ou 綿綢
min-hu 民戶
Ming 明
Ming-shih 明史
Ming T'ai-tsu 明太祖
mou 畝

Nagano 長野
Nan-hai 南海
Nan-hsun 南潯
Nan-t'ung 南通
Nanking (Nan-ching) 南京
nei 内
Nei chih-jan chü 内織染局
Nei-ko 内閣
Nei-wu fu 内務府
Ningpo (Ning-po) 寧波
Nung-hsueh hui 農學會

Ōshu 奥卅

pao (to guarantee, to secure) 保
pao (to reciprocate; retribution) 報
pao (to contract) 包
Pao-ch'ang ssu-ch'ang 寶昌絲廠
pao-chiao 包交
pao-hung 包烘
pao-kung 包工
pao-lan 包攬
pao-lan jen 包攬人
pao-shou 包收
pao-t'ou 包頭
Pan-yu 番禺
P'ing-hu 平湖
p'u-sa (bodhisattva) 菩薩

P'u-yuan chen 濮院鎮

San-shan hui-kuan 三山會館
san-shih 三市
san-shu 三叔
San-shui 三水
sha-tuan chuang 紗緞莊
Shameen (Sha-mien) 沙面
Shan-yin 山陰
shang-tzu 賞賜
Shang-yü 上虞
Shang-yuan 上元
Shang Yueh 尚鉞
Shanghai (Shang-hai) 上海
Shanghai chih-tsao chüan-ssu ku-
    fen kung-ssu 上海織造絹
                絲股份公司
Shanghai ts'ao-ssu-ch'ang
                上海繰絲廠
Shansi 山西
Shantung 山東
Shao-hsing 紹興
Shao-wen 昭文
shao-yeh 稍葉
she-shao 賒稍
Shen Lien-fong 沈聯芳
Shen Nung 神農
Shen Pao-chen 沈葆楨
Shen Ping-chen 沈炳震
Shen Ping-ch'eng 沈秉成
Shen-po t'ang 神帛堂
shen-tung 紳董
Sheng 嵊
Sheng-tse 盛澤
shih-ch'ih 市尺
shih-chin 市斤
Shih-lu 實錄
Shih-men 石門
shih-mou 市畝
Shinshu 信卅
Shōwa 昭和

Shuang-lin 雙林
shu-huo 熟貨
shui-t'ou 水頭
Shun-te 順德
Shuo-wen 說文
Soochow (Su-chou) 蘇州
ssu-chuang 絲莊
ssu hang 絲行
ssu-hao 絲號
Ssu-hui 四會
Ssu-pien kung-so 絲辮公所
Ssu-yeh kung-so 絲業公所
Su-ching 蘇經
Su-lun 蘇綸
Sung 宋
Sungkiang (Sung-chiang) 松江
Szechwan (Ssu-ch'uan) 四川

ta-pao 大包
ta-shu 大叔
ta-ts'an 大蠶
tai-chi chi-hu 代機機戶
T'ai-hu 太湖
T'ai-ts'ang 太倉
Taishō 大正
tan 擔
Tan-t'u 丹徒
Tan-yang 丹陽
T'an Ssu-t'ung 譚嗣同
T'ang 唐
t'ang-chang 堂長
Tao-kuang 道光
taotai (tao-t'ai) 道台
T'ao Pao-lien 陶保廉
Te-ch'ing 德清
ti (dry field) 地
ti (fabric) 綈
t'ien 田
Tōhoku 東北
Tokugawa Ieyasu 德川家康
Tomioka 富岡
Tonying (T'ung-yun) 通運

Toyotomi Hideyoshi 豊臣秀吉
Ts'ao-o 曹娥
Ts'ao Yin 曹寅
Tseng Kuo-fan 曾國藩
Tso Tsung-t'ang 左宗棠
ts'ou-chuang 縐莊
Tsung Yuan-han 宗源瀚
Tsungli Yamen 總理衙門
tsu-tsao 祖灶
t'u-hao lieh-shen 土豪劣紳
Tuan Fang 端方
Tung-wan 東莞
Tung-yang ching 東洋經
T'ung-chih 同治
T'ung-hsiang 桐鄉
T'ung-t'ing 洞庭
tzu-shu nü 自梳女

wai 外
Wan-li 萬歷
Wang Chia-liu 王駕六
Wang-chiang-ching 王江涇
Wang T'ao 王韜
Wang Ts'un-chih 汪存志
wei 緯
Wei-ch'eng 緯成
wen 文
Wen-chin kung-so 文錦公所
Wen-hsuan kung-so 文絢公所
Wu-chen 烏鎮
Wu-ch'eng 烏程
Wu-ch'ing 烏青
Wu-chün chi-yeh kung-so 吳郡機業公所
Wu Fu-hsiang 吳富鄉
Wu-hsing 吳興
Wu-k'ang 武康
Wusih (Wu-hsi) 無錫

Yamanashi 山梨
yang-ching 洋經
Yang-ch'ou 陽綢

275

yao-chi 腰機
yao-i 徭役
yeh-chi ko 野奚葛
yeh-sang 野桑
Yen-chou 嚴州
Yi Chang 怡章
ying-yeh shui 營業稅
Yokohama 橫浜
Yü-ch'ang 裕昌
Yü-ch'ien 於潛
yü-chien shang 餘繭商

Yü-hang 餘杭
Yü-yao 餘姚
yuan 圓
Yuan 元
Yuan-ning hui-kuan 元寧會館
Yun-chin kung-so 雲金公所
Yung-cheng 雍正
Yung T'ai 永泰

zaguri 座繰

Glossary II

Types of Silk Fabrics

The number and variety of silk fabrics woven in China, both in the heyday of silk and in more recent times, is truly astounding. The Maritime Customs survey of the silk industry, *Silk* (Shanghai, 1881), lists in its catalog of silk piece goods exported from Shanghai over 400 varieties of fabrics (pp. 86–100). This list represents only a portion of all the silks woven in China, primarily those of the Kiangnan area.

In general, silk fabrics in China were known as *ch'ou-tuan,* of which there were several major classes, as listed below. Since the names for these broad classes have been translated variously in English, I have preferred to use the transliterated Chinese terms in this study. For explanations of these Chinese terms, the reader is referred to the above-mentioned survey, and to D. K. Lieu, *The Silk Industry of China* (Shanghai, 1941), pp. 240–243, as well as translations given by E-tu Zen Sun and Shiou-chuan Sun in their translation of Sung Ying-hsing's *T'ien-kung k'ai-wu* (University Park, Pa., 1966), pp. 58–60, as well as *Mathews' Chinese-English Dictionary*, Rev. ed. (Cambridge, Mass., 1943).

chin 錦 . Mathews refers to this as thin brocade. Lieu says, "Fabrics of this group are made of multi-coloured silk."

ch'ou 綢 . Mathews says "thin silk." Lieu translates *ch'ou* as damask: "This group consists of the most widely sold plain or designed (not raised) silk fabrics." He includes *ts'ou* as a sub-category of *ch'ou.*

chüan 絹 . Mathews says, "a thin cheap silk; a kind of pongee." Lieu, "a kind of fabric made of inferior raw silk. It is light and thin." Sun also calls this a pongee.

jung 絨 . Velvet.

ling 綾 . Mathews says "damask or thin silk," while Lieu says, "fabrics of the same texture as *tuan* but . . . thinner and lighter in quality." Sun also says a damask.

lo 羅 . Mathews, "gauze, a thin kind of silk." Lieu, "This is also a thin and light fabric used as summer clothing material." Sun, p. 71, n. 11, notes that "gauze fabrics have long been known in China as *lo* and *sha*. The complex *lo* fabrics were much more popular than the simple *sha* fabrics in the Han and T'ang dynasties . . . In the Sung dynasty . . . however, the popularity of *lo* declined while the manufacture of *sha* was considerably increased. From this time onward *lo* was gradually replaced by *sha* and fell into disuse after the overthrow of the Ch'ing dynasty in 1912 chiefly because *lo* was more expensive than *sha* to manufacture."

sha 紗 . Generally translated gauze or thin gauze. Also for summer wear. See under *lo* above.

ts'ou 縐 . Generally translated as crepe, although *ch'ou* is also called crepe on occasion. According to Mathews this character should be pronounced *chou,* but I follow Lieu's practice of transliterating it *ts'ou,* which is more in conformity with Shanghai dialect pronounciation, in order to avoid confusion with the term *ch'ou.*

tuan 緞 . Satin, of many varieties.

Hua-ssu-ko 華絲葛. "A pure silk fabric with non-raised designs. It is finer than the variety known as *hutsou* [sic] and has a very good lustre. Its principal use is for clothing." *Hua-ssu-sha* is similar, "but finer and lighter in quality. It is usually white or dyed in light colors and used as summer clothing material." (Lieu, p. 243). This was a new style of silk fabric popular in the twentieth century.

Index

Harvard East Asian Monographs

STUDIES IN THE MODERNIZATION OF THE REPUBLIC OF KOREA: 1945–1975